E.A. BURTT, HISTORIAN AND PHILOSOPHER

BOSTON STUDIES IN THE PHILOSOPHY OF SCIENCE

VOLUME 226

E.A. BURTT, HISTORIAN AND PHILOSOPHER

A Study of the author of The Metaphysical Foundations of Modern Physical Science

by

DIANE DAVIS VILLEMAIRE

McGill University,
Montreal, Canada

SPRINGER-SCIENCE+BUSINESS MEDIA, B.V.

A C.I.P. Catalogue record for this book is available from the Library of Congress.

ISBN 978-90-481-5937-6 ISBN 978-94-017-1331-3 (eBook)
DOI 10.1007/978-94-017-1331-3

Printed on acid-free paper

For my Mother and Father

Elizabeth Galvin Davis (1916)
and Ellsworth Quinlan Davis (1916 – 1978)

In memory of their struggle to find meaning in life.

TABLE OF CONTENTS

PREFACE

I am grateful for the kind support and guidance of many good people during the course of my research and writing about the philosophy of E.A. Burtt. Professors Patrick Hutton and Alfred Andrea in the History Department at the University of Vermont pointed me onto the path of scholarship and have continued to hold a candle to light the way for me for the past ten years. Without their faith and encouragement this work would not have been undertaken.

At McGill, in Montreal, Canada, Professor Valentin Boss disciplined my thinking and writing while maintaining valuable enthusiasm for the research. From the first minutes of our initial meeting in 1994 right on through the finished product, Professor Boss has been my best critic and best inspiration.

At the foundation of the published work lies an allegiance to Burtt and his memory held deeply by so many who knew him. Friends of E.A. Burtt in Ithaca, New York offered me generous hours of conversation. Among the many who welcomed me and my questions, supported me with their encouragement and generosity during the time I spent in Ithaca and afterward, I want to thank especially Frank Pineo, who came upon me unexpectedly at the Burtt house one hot summer afternoon, greeted me with Quaker kindness, and has supported me with unwavering enthusiasm ever since. I warmly acknowledge and thank all the members of the Ithaca Society of Friends, who love to remember both "Ned" Burtt and his wife, Marjorie. They are Caroline Pineo, Helen and Karl Schantz, J. Anthony (Tony) Gaenslen, Peggy Walbridge and John and Carol Kimball. All were most kind and helpful.

Freddie Burtt Brinster, Burtt's daughter, has been a great listener and provider of good insight and balance on her father's intellectual and personal life. I thank her for her trust in me to handle her father's intellectual legacy with integrity and fairness.

I also want to thank Frank Moriarty of Australia, who shares an interest in Burtt, and Jean Matlock, who shares and interest in both Marjorie and Ned Burtt. Both of these good scholars provided me with copies of their own theses on the Burtts.

The response to inquiries I received from students at Cornell who had known Burtt as Professor Burtt was generous and helpful. I want to thank Professor David Braybrook and Professor Marcus Singer for their substantial help, also, Milton Konvitz and Elsie Stainton for their interest and contributions to my research.

I owe a lot to Professor H. Floris Cohen of The Netherlands, who has encouraged me in my interest in Burtt, both personally and with his own interest and commentary on Burtt's historiography of science in his book, *The Scientific Revolution.* Professor Cohen is the first scholar to have established Burtt's place among the great historians of science. His faith in my project made it possible to complete it and bring it into publication.

Any work of scholarship and writing makes demands on the family from which the life energy flows into the scholar. My daughter, Emily, and her father, Bernard, have been supportive and patient during the process and I appreciate that very much. Finally, Bernard undertook the tedious task of reading the entire work in order to offer comments and suggestions.

All of these good people have played a part in bringing a book on the life and work of E. A. Burtt into print. It has been a long and exciting process for me. E.A. Burtt has been an unduly neglected twentieth-century figure. I hope my attention to his work will be just a very small part of the growing interest in Burtt studies.

Diane Villemaire
Moretown, Vermont
July, 2001

INTRODUCTION

E.A. Burtt brought a pair of insoluble ideas into collision--metaphysics and science--beginning with the curious word juxtaposition in the title of *The Metaphysical Foundations of Modern Physical Science*. The work has a majestic strangeness, owing to the penetrating elegance with which Burtt poked into Newton's authority. Originally, *The Metaphysical Foundations* was Burtt's Ph.D. dissertation, written at Columbia University, which was the seat of American pragmatism and naturalism in the 1920s. It was titled *The Metaphysics of Sir Isaac Newton: An Essay on the Metaphysical Foundations of Modern Science.*[1]

For one thing, the work is a glimpse into a short, but exciting period in American intellectual history, a period noted for radical skepticism as well as progressive optimism. For another it is a barbed polemic against logical positivism and the new realism of Bertrand Russell, Alfred N. Whitehead, and the American New Realists, led by F.J.E. Woodbridge.[2] Most important and most enduring, if only enduringly controversial, it is philosophy of science, based on historical inquiry. While Burtt strikes some readers as "elegiac", others find him post-modern.[3] What is remarkable is that today at the end of the century, Burtt, both in and out of step with the times, is still refreshing.

Like it or not, *The Metaphysical Foundations of Modern Physical Science* continues to be read by undergraduates in colleges and universities everywhere. Burtt continues to have an effect on the intellectual lives of educated persons, whether they have come under his spell directly by reading *The Metaphysical Foundations* or indirectly by way of their professors and teachers who have saved a place, at least in their private speculations, for the kind of questions Burtt raised. Since 1924 when *The Metaphysical Foundations of Modern Physical Science* first appeared in England Americans have continued to flirt with its perspective, usually without being aware of the American Pragmatic tradition from which it sprung.

E. A. Burtt lived to be nearly one hundred years old, and he continued to write well into his nineties. Throughout his life, the core of his thinking remained unchanged, although he modified his methods and his interests ranged widely and shifted emphasis. Given the scope and impact of Burtt's thought, it is only natural to ask: Who might have shaped his thinking? Whom did he admire? With whom did he study? Who joined him in his free hours and what did they discuss? What forces shaped E. A. Burtt's thinking during the genesis of *The Metaphysical Foundations of Modern Physical*

Science and how did those influences wax and wane over the course of his life as a philosopher?

What were the radical ideas filling the air in New York City around Columbia University, City College, and Union Theological Seminary when Edwin Arthur Burtt came there to study in 1915? What were the unique interests of American philosophers in those days and what special tensions shaped that fitful era, just before the first Great War?

Among the many bold assertions Burtt made concerning early modern science the key to all of them was his demonstration that the principle of universal order, upon which the whole edifice of science knowledge is based, is metaphysical.

Burtt's research for his dissertation involved the careful study all of Newton's writing. Unlike any previous historian, he accepted the large volume of religious expositions right along with the scientific. Even in 1924, there was plenty of evidence to suggest that Newton had theological interests. Other historians had simply not found them important to their research questions or worse, had found them embarrassing to Newton's reputation as a scientist and intentionally ignored them. Burtt studied Newton's theology right along with his science and reached the revolutionary conclusion that Isaac Newton "agreed with [Henry] More on the extension of God...and he certainly believed in extended ethereal spirits."[4] This is the "metaphysical foundation" upon which Newton's science rests, and it is the only foundation we have for our belief in universal order.

What is more, for Newton, the connection between the knowing agency locked in the brain and the material world to be known was bridged by "a spiritual continuity connecting all the links in the infinite scene" and it was "supplied in God." The problems of knowledge, which Newton resolved by invoking the presence of God, evolved and changed as Newton's philosophy was popularized by Voltaire and the other Enlightenment thinkers. While Newton had evaded some philosophical problems by appeal to God, Hume, Locke and Kant pried off the Deity, leaving the situation with logical inconsistencies. No questions were raised, however, since the whole thing rested on the unquestioned authority of Newton, which had replaced God in the mind of modern man. For Burtt, it is one of the ironies of history that although it was not philosophically consistent with Newton's own beliefs, the "general picture of the universe and of man's place in it which went forth under his name was essentially the one [worked out by Galileo and Descartes] in its most ambiguous and least construable form."[5]

Burtt argued that since the principle of universal order, upon which modern physical science is based, is a metaphysical principle it is not necessarily a structural principle of the universe itself. According to Burtt the scientific world-view which has pervaded our common sense understanding of ourselves in the universe is nothing more than a transient episode in the history of man. In the 1932 revised edition to *The Metaphysical Foundations* he suggested with much modesty the need for a new epistemology and a new world-view, but he laid no claim to such a system himself. What he recommended was the critical historical analysis of the basic presuppositions of the world's major idea systems, including modern science, the various schools of modern philosophy, and world religions, as a necessary first step.

Although he denied as much in his unassuming questioning, the fact is that E.A. Burtt was working on a new epistemology of his own. It is known only from a brief manuscript, *The Genesis of Hypothesis,* which was abandoned sometime after 1935. This manuscript makes a better understanding of *The Metaphysical Foundations* possible.

Burtt had been profoundly influenced by John Dewey. One of his early teaching assignments at Columbia University had been a course in *Reflective Thinking*, which was an analysis of the process of problem solving, based almost entirely on Dewey's reconstruction in philosophy. Dewey's logic diverged widely from accepted doctrine since he rejected the truth-falsity categorizations of propositions, and held instead that propositions were simple tools in the process of inquiry, no more true or false than a hammer or screwdriver is true or false.

While Burtt acquired the essentials of Dewey's thinking through teaching the Deweyan methods of inquiry at Columbia, he displayed a tremendously creative freedom in using those methods to examine the history of science from a new perspective. Obviously if scientific propositions are neither true nor false, but only tools in the process of inquiry, the notion that science is a cumulative enterprise heading toward certain correspondence with one, true physical reality is undermined and important questions about scientific truth can be raised.

There are parallels between this type of reflection on the progress of science and Thomas Kuhn's historical basis for rejecting the strict empiricists' interpretations of the logic of science. Kuhn, however, recalls that he did not read Burtt's *Metaphysical Foundations* as philosophy of science, finding it useful simply as history.[6] The inspirations Kuhn does acknowledge are Herbert Butterfield and Alexandre Koyré.

Even so, ideas parallel to Burtt's are liberally threaded through the work of Herbert Butterfield and Alexandre Koyré as well as through that of Thomas Kuhn. Henry Guerlac, professor of the history of science at Cornell during Burtt's time there, has written that Alexandre Koyré, in personal conversation with him "once remarked that his reading of E.A. Burtt's remarkable book, the *Metaphysical Foundations of Modern Physical Science*, played an essential role "in leading Koyré to what he has characterized as his 'conversion' from the history of religion to the history of science."[7] In addition, Pietro Redondi, in the preface to *Alexandre Koyré, De la mystique a la science,* has cited Koyré and written:

> Koyré confimera dans les Etudes galiléennes son adhésion à ce grand livre: "C'est M. Burtt qui nous paraît avoir le mieux compris la substructure métaphysique-mathématisme platonisant - de la science classique."

Redondi has substantiated Burtt's influence on Koyré several times.[8] Butterfield and Kuhn were both influenced by A. Koyré[9] and since Koyré's work has a debt to Burtt, Kuhn has built on Burtt's shoulders whether or not consciously.

For example, Burtt's speculations in the conclusion to the 1932 edition of *The Metaphysical Foundations* parallel ideas Kuhn has advanced.

> How to construe a rational structure out of...nature is the great difficulty of contemporary cosmology...[The] difficulties suggest that perhaps we need to be much more radical in the explanatory hypotheses considered than we have allowed ourselves to be heretofore. Possibly the world of external facts is much more fertile and plastic than we have ventured to suppose; it may be that all these cosmologies and many more analyses and classifications are genuine ways of arranging what nature offers to our understanding, and that the main condition determining our selection between them is something in us rather than something in the external world. This possibility might be enormously clarified by historical studies aiming to ferret out the fundamental motives and other human factors involved in each of these characteristic analyses as it appeared and to make what headway seemed feasible at evaluating them, discovering which are of more enduring significance and why.[10]

Kuhn's thought has raised no small amount of controversy, with Kuhn himself adding to the fray by advancing and then retracting ideas. One way to pass over the confusion is to accept Kuhn's thought through the eyes of Paul Hoyningen-Huene. Paul Hoyningen-Huene spent a year at MIT with Kuhn in preparation for his own exegesis of Kuhn's philosophy of science. The outcome of Hoyningen-Huene's year is his book, *Reconstructing Scientific Revolutions*. Kuhn, as interpreted by Hoyningen-Huene, was concerned with questions raised by Burtt. Ideas put forward in the chapter titled, *The Constitution of a Phenomenal World*, compare substantially to Burtt's examination of *something in us rather than something in the external world.* Hoyningen-Huene wrote, "The suggestion that perception *in general* might be influenced by subject-sided factors...becomes plausible, along with the corollary that the subject-sided factors might be systematically different in different cultures, being the product of systematically different learning processes. It follows that the phenomenal worlds of two different cultures can be different, in which case the empirical concepts characteristic of one will find no precise counterparts in the other."[11]

Hoyningen-Huene, standing in for Kuhn, confronts the fundamental question. Can we know the real world, the world as it is in itself? And how could we ever know that we know it, since we can have no objective model with which to compare our understanding. Given that humans know only the "phenomenal worlds" how can universality be claimed for any one of those?

> How might we show that the general analysis of the constitution of phenomenal worlds actually generates results which hold for *all* phenomenal worlds? The most appealing strategy involves confirming the analysis in concrete individual cases. But even if this confirmation is successful for all past phenomenal worlds known to the historian, no real claim to universality can be asserted, for nothing ensures that the discovered factors in the constitution of this class of phenomenal worlds really apply to all phenomenal worlds. It might be that phenomenal worlds in the examined class were all constructed in comparable ways because they all belong to a common, more or less narrow tradition....
>
> An approach that takes the anthropological basis of world constitution into account appears to offer the only prospect for a more rigorous justification of claims to universality. This approach requires that we know which capacities needed for the constitution of a phenomenal world humans come equipped with and which are developed over the course of interaction with the environment. If our knowledge of the anthropological foundations of world constitution is solid enough, we will presumably be able to make universal claims about the processes and structures which must necessarily participate in world constitution.[12]

Paul Hoyningen-Huene's interpretation of Kuhn, if not Kuhn's own thinking, echoes Burtt's quest for a new philosophy of mind, taking critical historical analysis into account in order to arrive at as clear a picture of physical reality as it is possible for humankind to achieve.[13]

The fact that *The Metaphysical Foundations of Modern Physical Science: A Historical and Critical Essay* has proved to be a title with enduring appeal turns out to be no mystery. Indeed, it is a book which has led countless readers to question the foundations of scientific knowledge and has opened doors to new ways of thinking about science, philosophy, and what we can hope to know about objective reality in the present age. The book is a classic because it brings us back to unanswered questions.

This was Burtt's intention. He concluded his revised edition with the hope that one of his readers might "catch some glimmerings" from his probing questions and thereby contribute to the development of "an adequate philosophy of mind," deemed by him to be the essential first step in the construction of a new cosmology. "For myself," he wrote, "I must admit that, as yet, it is beyond me, and only insist that whatever may turn out to be the solution, an indispensable part of its foundation will be clear historical insight into the antecedents of our present thought-world."[14]

Until now, Burtt has been more-or-less ignored as an influential American author confronting these questions. Skepticism, while part of the profile of the American mind, does not easily turn itself against science, the bed-rock foundation of the U.S. intellectual edifice, the progenitor of its great wealth and world influence. The American Naturalists who influenced Burtt admired and used scientific methods, wanting to turn them on as search-lights in all fields of human endeavor. With piercing insight, Burtt turned some of the most potent philosophy of the Scientific Revolution, David Hume's skepticism, back onto science itself with some unsettling, and still unresolved, questions about the logical foundations of science.

Is Thomas Kuhn in a better position to ask these same questions in the late twentieth century than Burtt was in his own era? If he is, it is owing in part to Kuhn's pedigree as a physicist first and to the popularization of Burtt's ideas by the Frenchman, Alexandre Koyré. Both these facts, bolstered by post-modernism, imported from Europe, helped to prepare the ground for Kuhn's ideas. American thinkers have shied away from skepticism about science while at the same time being unsettled by science's rigidity, and this has made *The Metaphysical Foundations* both the rose and the thorn on university reading lists.

CHAPTER 1

COLUMBIA UNIVERSITY IN THE 1920S: THE YOUNG RADICAL PHILOSOPHERS

E.A. Burtt belongs in the tradition of Pragmatism and Naturalism, which developed in the United States in the early twentieth century. The so-called golden age of American philosophy was marked by the belief that philosophy was the guide to life and that a proper philosophical approach would yield answers to social, political, ethical, and religious questions. Science was taken for granted as an important key to the future, although some Naturalists with a more idealistic bent were concerned about the negative effects science could have on the human spirit.

The center for pragmatic naturalism during the 1920s was Columbia University in New York City. Ideas which go under the heading of Pragmatism have diverse sources but, are generally associated with Charles S. Peirce, William James, John Dewey, and, later, George Herbert Mead. Peirce and James were classmates at Harvard and continued as friends all their lives. Mead and Dewey first met as colleagues at the University of Michigan and were colleagues again at the University of Chicago. A lesser-known group of Pragmatic Naturalists included the young radicals around Columbia University in the 1920s, mostly disciples of John Dewey. Although unacknowledged by Burtt, it is Dewey's reconstructed philosophy, to be explained below as genetic history and reflective thinking, which stands behind *The Metaphysical Foundations of Modern Physical Science*.

Pragmatic naturalism grew out of two more or less conflicting trends in American culture, a religious world-view coupled with a decidedly practical attitude toward the concerns of daily life. Concerns about reconciling the here and now with speculation about the hereafter generated an anti-philosophic, anti-intellectualist trend among American academics. The naturalists, inspired more by Charles Darwin than by Isaac Newton, turned away from the supernatural and toward the natural and material. They abandoned the dogma of Calvinism, but not the Puritan emphasis on right living and thus became practical moralists, concerned with ethical behavior which could square itself with free-thinking, individualism, and principles of the American Enlightenment, a belief in the power of faith and the power of reason.

American naturalism holds that whatever there might be is entirely an aspect of nature and can be approached scientifically. Twentieth-century American naturalism has tended to hold on to the Spinozist conception that nature includes more than simply matter in motion, and whatever such nonmaterial aspects of nature there might be, they are not reducible to matter nor explainable in material terms alone. According to John Ryder in the introduction to *American Philosophic Naturalism* two traditions developed out of this basic belief. One is that although nature consists of both material and non-material phenomena, matter is the more fundamental on the basis of the fact that anything that is non material emerges from the material. Ryder attributes this view to Roy Wood Sellars, John Dewey, Ernest Nagel and Sidney Hook. Dewey, Nagel and Hook were connected with Columbia University during Burtt's years there (1916-1923).

The other view, Ryder attributes to John Herman Randall, Jr. and Justus Buchler, both of them at Columbia in Burtt's day. This is what might be called metaphysical pluralism. It holds that there is more to be found in nature than simply matter, and it refuses to acknowledge primacy to any aspect of nature. As a close friend and colleague of John Herman Randall, Jr., Burtt is allied with this second view.

The influence of natural science, especially Darwinian evolution and American Enlightenment "free-thinking," have been currents which blend easily with elements found in Spinoza and the philosophy of David Hume. Spinoza rejected Descartes' distinction between material and mental substances and proposed, instead, the existence of a single substance which he described interchangeably as God and Nature or thought and extension. In this way Spinoza made the natural and the supernatural indistinguishable; all is in nature and all is in God. Nothing, therefore, is outside of nature. By treating matter and mind as alternating parts of the same thing in nature, Spinoza asserted the continuity of the mental and the physical, dissolving Descartes' dilemma.

If all is nature and we have scientific methods for inquiry into nature, we have, or should be able to develop, methods for inquiry into the processes through which mind knows nature and then knows itself. Burtt was impressed by the possibilities here, and he continued read Spinoza, finding new meaning there, all of his life.

The naturalists were anti-reductionist, and unlike Spinoza, anti-determinist. They regarded the world as being open to chance and randomness as well as describable by principles or laws of nature. Generally, the Naturalists reject Cartesian dualism and emphasize experience as continuity between human beings and nature. The implications for our understanding of nature, as it is in itself, objectively, apart from us, but through our experience of it alone, presents unresolved questions. Burtt intended to reconcile some of these questions through his proposal of a new philosophy of mind. He hoped to work out a philosophy to explain our experience of nature without making any claims for the correspondence between that experience and objective reality, whatever that might be.

In 1935, he was working on a new theory of perception and cognition, which he called a science of metaphysics or theory of metaphysical categories. The manuscript, called *The Genesis of Hypothesis*, was never completed.

Ever since Descartes and Newton the project of modern philosophy has been to work out how we can understand an objective and absolute reality, wholly independent of mind. Borrowing from the scientific method, the philosophical enterprise has been concerned with the discovery of that reality, with what is truly the case as opposed to what merely appears to be the case. Here lies the central problem of modern philosophy. How is this to be done? Is knowledge of objective reality possible? How is it possible? With respect to material nature, the answer has been to apply the methods of science. But with respect to other less tangible matters philosophers have generally turned to one or another of the two main foundations of knowledge, reason or simple sense perceptions. Most of American twentieth-century philosophizing, especially since the 1930s, has been a variety of technical distillations of seventeenth and eighteenth-century ideas. At the bottom of these approaches in whatever form or mix they appear is the basic presupposition that an objective reality exists and can be discovered and known by mind. This, of course, is the position bequeathed to us by modern science, as we conceived the legacy from Descartes, Newton and Hume.

This is what Burtt wanted to put to the question in *The Metaphysical Foundations of Modern Physical Science.* Here is where contemporary postmodern philosophy has a strong Burttian flavor. To put it simply, modern philosophy based as it has been on modern science has presupposed a truth independent of the inquirer and unaffected by the process of inquiry. Developments in science itself and in the recent philosophy of science challenge these presuppositions. It is now well known, at least in some instances, that the process of scientific study affects the objects being studied. In present-day philosophy of science the former belief that science methods are leading us progressively toward certain truth about the natural world is, at least, questionable. The idea that we can get a pure god's eye point of view has been discounted by a number of respected contemporary thinkers.

A useful postmodern insight is that inquiry is always conducted from some point of view, from a particular perspective, which we bring with us. We invariably approach a problem with a set of ideas and concepts which have helped us to identify the problem in the first place. In one way or another we always operate our minds with some purpose in mind or with some interests which drive our finding the solution. We must come at the problem from some perspective. We can not find an objective point of view from which to attack the problem of finding an objective reality because this would be the equivalent of finding a perspective which is not a perspective.

Arguably, many thinkers have been prepared for this insight by E. A. Burtt. Burtt said it precisely: thinking without a perspective (Burtt called it metaphysics) is impossible. In one way or another, this is the challenge to modernity which runs through all the post-moderns from Foucault and the post-structuralists and from the pragmatists, like Burtt, to Richard Rorty and his antifoundationalism. There is, undeniably, a frightening implication in all of this--objective inquiry and consequently objective knowledge is impossible. The unsavory postmodern position which makes Burtt the enemy of the modernists and mathematical realists is that either nature is continuously constructed by human activity, or that whatever objective reality there is, we have not got the inquiry methods to know it, and it is therefore pointless to concern ourselves with it.

This is a position which has been attributed to John Dewey, but Dewey did not make much of physical reality or philosophy of science either. It is also a position that served the early twentieth-century American philosophers' fundamentally idealist motivations and religious concerns. They were at odds with modern science and its promises for truths outside of Absolute Truths, religiously revealed. Some of them, like Burtt, sought a new synthesis and reconciliation between science and religion. Their program required devaluation of science to achieve renewed value for religion. Burtt's early arguments were sometimes intended to put scientific and religious ways of knowing on an equal footing.

It was 1915 when E. A. Burtt moved to New York City and enrolled at Union Theological Seminary with the intention of preparing for the ministry to follow in the footsteps of his father and grandfather. In those days Union had an agreement with Columbia University that allowed for cross enrollments and he signed up for some courses there, too. At Columbia, Burtt was in the company of the intellectual giants of the age. F. J. E.Woodbridge, John Dewey, James Harvey Robinson, and Charles Beard. All were great names in the Golden Age of American Philosophy, a time when philosophy was the activity of college professors who performed a kind of ministerial function in cultured society. In the style of Emerson and the Unitarian clergy, these men wrestled with moral and religious problems relevant to American public life. Burtt found a niche among these thinkers, and, in time, he abandoned his early career goal in favor of pursuing the life of a university professor.

During the early part of the twentieth century an axis extended from Columbia to the University of Chicago along which the pragmatists and the new realists were engaged in vigorous debate on a number of problems, such as the relationship between reason and nature, the distinction between mind and matter, the problem of respecting the spirit of science and the spiritual interests of humanity. On Dewey's head of steam they lectured and wrote in anticipation of radical changes in society to be brought about through general education. It was the period of America's great industrial expansion and the philosophers were trying to fashion the instruments for a moral criticism of their age. The level of discourse, in both its intensity and its range, was never before paralleled in America; nor has it been recovered since, in the sense that the topics up for philosophical discussion were accessible and intelligible to the ordinary educated person.

Pragmatic naturalism was more of a movement than a systematic philosophy. During the years that Burtt was associated with Columbia, the leading ideas and methods of the movement were being analyzed and developed by its critics and followers. At Columbia the analysis was carried out by reflective thinking, a formal process recommended by John Dewey, which begins with specific doubt. The specific doubt was not universal doubt as Descartes had begun his philosophy because the pragmatic naturalists did not think that it was psychologically possible to doubt everything at once. They believed that intellectual activity can not take place unless some beliefs are taken for granted. The task is to know just what is taken for granted in order to put it to the test at some future time.

Pragmatic naturalism is really a problem solving method which takes its strength from treating problems within the context of specific, well-defined situations. All the same, the pragmatic naturalists were concerned with the problem of the unity of knowledge and all believed that scientific knowledge must be brought into harmony with other beliefs about the world, including values. All these themes were drawn together by E. A. Burtt in his analysis of the Scientific Revolution and perpetuated in his later thinking.

In the preface to *The Metaphysical Foundations* Burtt acknowledged F.J.E. Woodbridge, Morris R. Cohen and John Herman Randall, Jr. for their contributions to his work. Woodbridge, who was Chair of the Philosophy Department at Columbia, became Burtt's thesis advisor. But there were other influences, not mentioned in the preface. John Dewey's Reflective Thinking and James Harvey Robinson's New History are central to the development of Burtt's perspective. No one studying and teaching at Columbia University in the 1920s could have escaped these powerful influences.

The Metaphysical Foundations of Modern Physical Science reflects a point of view deep-rooted within this special brand of philosophical Naturalism dominate at Columbia University in the 1920s. John Dewey and his followers, believed that science was powerfully shaping American culture and they wanted to study it as a cultural phenomenon. They proposed the idea that science should be understood within the context of modern culture.

Burtt chose the genetic method of history to examine the contemporary problem. As deeply as *The Metaphysical Foundations* penetrates into the history of science, it penetrates into the philosophy of science and into the philosophy of history. In this it is a rich amalgam of some of the most important, and certainly the most original, ideas ever to have sprung out of American academia. The questions present-day philosophers of science are asking about context-dependent theories of physical reality grew out of the Pragmatists' interest in analyzing science within the context of the culture which values and produces it.

F.J.E. Woodbridge and Morris R. Cohen tended toward a realist outlook, more scientific, more interested in historical detail and even more disciplined in their logic than the more idealist types, Dewey and Robinson. Woodbridge had trained Cohen in the history of philosophy. Dewey and Robinson were more interested in social change. Burtt fell somewhere between them. Playing his role as naysayer and devil's advocate, Morris R. Cohen derided what looked like context-dependent science as "the theory that science deals in 'fictions'."[1]

Cohen, perhaps the most famous teacher of philosophy ever to work with undergraduates in the United States, was not an official part of the Columbia establishment. Still, his influence was unmistakably there. Sometime during the preparation of his dissertation, Burtt met him. Cohen had been appointed Assistant Professor of Philosophy at New York City College in 1912. Prior to that he had taught mathematics in high school and later, at City College. All the while, he had been anxious to break into the field of Philosophy in which he had earned his Ph.D. at Columbia with credits in courses taught by F. J. E. Woodbridge, Felix Adler and W. H. Sheldon. By the time Burtt met him, Cohen was well-known and highly-respected as a naturalist philosopher.

Morris Cohen's presence within the Columbia philosophy department was dynamic and sustained, even as his hopes to secure an appointment there were thwarted by Woodbridge and the university administration. Apparently, this was owing to the fact that Cohen was an immigrant Jew and that philosophy in U.S. universities and colleges was typically taught by the college president, who would also be a Doctor of Divinity and the college chaplain. While it seemed to be out of the question that Cohen would teach classes at Columbia since Jews did not typically get such appointments at American colleges until after World War II, his influence at the school was profound. Many of Cohen's former students from City College were enrolled there and they continued to seek out the company of their provocative former teacher. He was a powerhouse, whose combative wit, acrid humor and love of argument were relished by many young men who vied for his attention.

According to a recent account by Marcus Singer, Cohen was a remarkable teacher and well known throughout New York City. His best students were poor, bright, eager and hard-working immigrant boys who went tuition-free to City College where "the student body rather than the faculty created the intellectual tone."[2] According to Singer "[Cohen] was thought of, in the informed public mind that knew something about intellectual and cultural matters, alongside Dewey and Russell as one of the most famous philosophers of the day."[3]

Cohen prepared an extraordinary number of undergraduates at City College who later became prominent philosophers.[4] It was natural that a young man like Burtt was attracted to him.

The rest of the younger generation at Columbia in Burtt's day included John Herman Randall, Jr., Randolph Bourne (died December, 1919), James Gutmann, and Harry Elmer Barnes. Later additions to the list were Sidney Hook and Ernst Nagel, two of Cohen's boys from City College. They were the so-called young radicals, admirers of Walter Lippmann, Herbert Croly, *The New Republic*, and *The Nation.* Cohen authored articles for both papers. They believed in progress and in science, but they were not mechanistic or deterministic where the human mind was concerned. Throwing off the notion that minds can do no more than reflect the order of nature, they believed in the power of faith and above all the power of human reason. They denied the Absolute, but they were moralists. That is, they were interested in human ideas about ethics and morals and they believed that philosophy, properly practiced, should illuminate and guide all other human endeavor--science, law, history, economics and religion. They took as their province new political and social thought.

During the Golden Age a radical reappraisal of formal traditions in every field was underway, guided mostly by Dewey's faith in human intelligence as the agency of human liberation and the conviction that what people want most of all is freedom, defined as the power to think for themselves. The expectation that society would be reformed based on revived enlightenment ideals and the sheer excitement of being on the cutting edge of an important new era held the young radicals together in spite of their varied philosophical temperaments. History has demonstrated that they were primarily optimistic theorists, after all, and not practical reformers themselves.[5] Contributions to educational theory and the writing of textbooks were their greatest accomplishments. None produced tightly argued, rigorous tracts of substance as did the later analytic philosophers who took over the field after the second war. Furthermore, because they emphasized history, their innovations and philosophical critiques were buried in narrative historical accounts. All are mostly forgotten today. None of the young radicals, with the possible exception of Sidney Hook, who became a communist for a time, would ever achieve the influence and stature of their teachers, Dewey, Woodbridge and Cohen. Even Dewey is long out of fashion.

By the time the youngest of the young radicals came into the power of their own careers, in the 1930s and 1940s, academic fashions had shifted. The social and moral questions which had interested them were swept aside. Their much-vaunted method, historical analysis, was slandered as "Hegelian" and many of them were dubbed "antiquated" before their own philosophical thought matured. By 1930, philosophy was a purely professional affair in the United States. Successful philosophers specialized within technical areas and looked with suspicion on those who popularized or tried to apply their ideas to the real world of society and culture. Burtt's *Metaphysical Foundations* has survived the best and longest of any of the work produced by the younger generation of Columbia radicals.

> To give meaning to the present, history must be treated as an ever-continuous genetic development in which the present is illuminated by perceiving anachronisms or the often irrational origins of our ideas and institutions.[11]

Such an approach combined well with an education in *Reflective Thinking*. Taken together they were intended to discipline the imagination and to make a new kind of intellectual history the cornerstone of a reconstructed philosophy. More than just the examination of logical idea sets from the past, in Dewey's and Robinson's program, intellectual history became a parade of human solutions to temporal problems of experience. Of course, this is exactly the line Burtt followed in his examination of early modern science. For Burtt the "logico-mathematical world-view" was just one in a passing parade of world-views, not more significant than others. At the end of *The Metaphysical Foundations*, Burtt appealed to his intelligent readers to glimpse a new world-view which could be permanent. In his later philosophy, the permanent world-view is what he came to call a philosophy of man.

It is unlikely that Burtt had any exposure to the genetic method before he came to New York from Yale and began taking classes at Columbia. As he assimilated the reformers' perspective, he applied it eagerly to his own research interest. After six years in association with Columbia Philosophy, Burtt had proved his mettle, and he was appointed a full-time instructor in 1921. Burtt joined a prestigious group of young intellectuals who were taken into the Department of Philosophy as teachers. They called themselves The Columbia Associates and were expected to design courses themselves. Simultaneously he worked on what was to become *The Metaphysical Foundations of Modern Physical Science.*

At the time Burtt was made an instructor, the Philosophy Department was being expanded rapidly to put it on a competitive footing with Harvard and the Sage School at Cornell. Consequently, selections into the department were made with utmost consideration for the future. The other university in the running for lead position in the field was the University of Chicago, about equal with Columbia, and just behind Harvard and Cornell in prestige.[12]

John Herman Randall, Jr. was the most well liked, most respected and most powerful member of this elite group of young men at Columbia and as historian for the department, he wrote the following account.

> The new members added during the expanding twenties all took an active part in the College program; there has never emerged the slightest division in the Department of Philosophy between a college and a graduate staff...The young instructors were from the start encouraged...to offer graduate courses in fields they could work up--originally historical, but later analytic or systematic as well....Woodbridge and Dewey were not averse to seeing established a band of

> younger teachers who would understand their lifework, be not unsympathetic to it, and perhaps even extend it in their own ways.[13]

Burtt expanded Woodbridge's interest in the history of philosophy by developing an advanced course on the British empiricists from which his textbook, *The British Empricists from Bacon to Mill*, evolved.

According to Randall, all the young men were John Dewey's disciples. The other senior faculty were: F.J.E. Woodbridge and Wendell T. Bush. Randall described the three of them this way:

> All three were in the vanguard of the pragmatic movement. And the younger men added to the rapidly expanding Department after 1918 had all been his [Dewey's] students: personal loyalty intensified intellectual conviction. They had been trained by Woodbridge and absorbed much from him; but in the end they agreed fundamentally with Dewey, even when they arrived by a wholly different route.[14]

Burtt was one who arrived by a "different route." In a biographical paper written for *Philosophy East and West* (1972) Burtt refused the tag "disciple" while admitting Dewey's influence on him. He wrote this account of himself:

> With what philosophical point of view did I begin my career as teacher? Looking back to that period, I see that it was a rather inchoate form of idealism, reflecting the liberal Protestant orientation I had then adopted. This idealism was gradually revised to make room for major aspects of Deweyan instrumentalism. In my years at Columbia, Dewey had impressed me as a person, and during my decade in Chicago, in close contact with colleagues whom he had deeply influenced, I came to accept what appeared to be the core of his contribution to philosophy. But continued sensitivity to insights from various quarters prevented me from becoming a disciple of any one philosopher or philosophical school.[15]

Whether Burtt realized it consciously or not, there was no conceivable way he could have escaped John Dewey's influence at Columbia. Although Dewey was lecturing in Tokyo, Peking, and Nanking between 1919-1921, the substance of those lectures, now published as *Types of Thinking Including, A Survey of Greek Thought,*[16] was prepared before the trip and was being used by a cadre of young instructors, including Burtt, in introductory classes in logic. The teaching group was led by John H. Randall, Jr., and they called themselves The Columbia Associates in Philosophy. Their course was titled, *Reflective Thinking as a Method of Problem Solving.* Burtt taught this course several times between 1921-1923. At first the Deweyan doctrines were taught from typewritten copies of a text especially prepared for the students by Randall and eight of the other young instructors, although Burtt was not one of them.[17] After a while, the course materials were published in a textbook titled, *An*

Introduction to Reflective Thinking. The book was used at Columbia and in many other universities. *An Introduction to Reflective Thinking* is basically an expanded spin-off from Dewey's *Studies in Logic* (1903) and *Reconstruction in Philosophy* (1916). Another of Dewey's books, *How We Think* (1903), was also used in the course.[18]

As Burtt taught from the Deweyan materials and interacted with the nine authors, especially Randall, he became a Deweyan-style philosopher, if only indirectly. Dewey's mark on *The Metaphysics of Sir Isaac Newton* is unmistakable. Sure enough, years later, while he was teaching at Chicago, Burtt published his own version of *An Introduction to Reflective Thinking*, calling it, *Principles and Problems of Right Thinking* (1928 and 1931). A revised edition appeared in 1948, titled simply, *Right Thinking.* The Preface credits the Columbia Associates and numerous citations trace the extensive debt to the Columbia teaching materials.

JOHN HERMAN RANDALL, JR.

Burtt's closest friend from his New York days was the academically energetic, Columbia favorite, John Herman Randall, Jr. Randall was born in Grand Rapids, Michigan on February 14, 1899, making him younger than Burtt by six and a half years. Randall's father, John Herman Randall, Sr., had been a distinguished intellectual historian who taught at Columbia during the early decades of the twentieth century. He was also a minister at Mt. Morris Church in New York City. John Randall, Jr. earned a B.A. from Columbia in philosophy in 1918; he took his M.A. in 1919, and his Ph.D. in 1922. Burtt began graduate studies in the same year as Randall began his undergraduate work, 1915. Randall was hired as an instructor in philosophy in 1920; Burtt was made an instructor in 1921. Randall was made a full professor in 1935 by which time Burtt had moved to Chicago, ended his career there, and moved on again to Cornell. There may be a hint as to why Burtt did not stay at Columbia in the fact that Randall had to wait so long for a professor's position. If Columbia was slow to offer appointments during the time philosophy was becoming professionalized, there were plenty of other jobs around. Burtt had a young family of four daughters to support and was already an older man, 33 years old in 1925 when he took the position at Chicago.

There were natural sympathies between E.A. Burtt, John Randall, Jr. and Randall's father, since all three shared an interest in religion. Burtt had been resentful of his own father's fundamentalist Christianity and the lively intellectual exchange of ideas characteristic of liberal Protestantism gave him the freedom to pursue his other interests in philosophy and science. The two young men collaborated on a book, *Religion in the Age of Science* (1929), and later supported religious humanism publicly by signing the *Humanist Manifesto* in 1933.

At first, Randall's interests were concentrated on intellectual history and the cultural context of philosophy, rather than science. The first line in his *The Making of the Modern Mind, A Survey of the Intellectual Background of the Present Age* states the historian's aim as "entering sympathetically into the spirit of the past, to make the thought of the present more intelligible." The title of his second book, *Our Changing Civilization, How Science and the Machine Are Reconstructing Modern Life,* illustrates his philosophical interests. In the last years of the 1920s, Randall wanted to address specifically the problems posed to civilization by the latest scientific developments. For him, as for Burtt, science was a problem for philosophers to analyze in order to help ordinary people synthesize it meaningfully into their lives. For Randall, as for Burtt, philosophy was a critique, mostly from an historical perspective, of the fundamental assumptions of mankind's various enterprises, science, philosophy and especially religion. In this both Burtt and Randall followed John Dewey. Randall wrote:

> [Philosophical problems are] those questions...which emerge... whenever strife of ideas and experiences forces men back to the fundamental assumptions in any field; when it compels men to analyze them, to clarify, criticize, and reconstruct them.[19]

As philosophy in America settled down to the business of testing the logical conditions under which this or that predicate is true or false as applied to a given subject in the manner of the scientific philosophers from Europe, the practice of philosophy as an interrogative art fell out of fashion and with its neglect and lack of exercise a work like *The Metaphysical Foundations* can take on a post-modern flavour.

At the time Randall was working on his sweeping intellectual history, *The Making of the Modern Mind* in 1921-1923 Burtt was working on his dissertation. It was through Randall that Burtt came under the indirect influence of James Harvey Robinson. As an undergraduate, Randall had taken classes with Robinson and with Charles Beard. Both men had left Columbia were teaching at the New School for Social Research by 1919.[20]

In 1924, the same year that Burtt published *The Metaphysical Foundations*, Randall published *The Western Mind: Its Origins and Development*. This book grew out of a course Randall had been teaching, *Introduction to Contemporary Civilization*. It was the successor to Robinson's course, *The Intellectual History of Western Europe*. Randall had taken the course over from Robinson when Robinson moved to the New School. Within two years, Randall had revised his book, *The Western Mind*, and published it under a new title, *The Making of the Modern Mind*. This book is considered to be Randall's companion to Robinson's *The Mind in the Making*.

The Making of the Modern Mind has a network of connections with Burtt's *Metaphysical Foundations*, testifying to the intellectual collaborations that had gone on. The chapter on Newton has numerous citations of Burtt's work and Randall rather obviously shared Burtt's perspective on the history of science.

Robinson's *The Mind in the Making*, Burtt's *The Metaphysical Foundations* and Randall's *The Making of the Modern Mind* were all three, written from a similar perspective on the meaning and methods of history. All three depend on the same pragmatic method of historical analysis, the so-called genetic method of interpreting history to make our experience in the present meaningful.

Randall once reflected on his youthful sense of purpose, absent any special plan for accomplishing it, in an essay on historical naturalism. The outlook he describes applies equally well to his friend, E.A. Burtt. The two of them began their careers like a pair of brothers.

> I remember, as a college student, being asked by a friend the not unusual question, what did I intend to do? I can still recall the mood in which I realized I could not answer. Such a question had never seriously occurred to me. "Why," I replied, "I really do not know. But I think I do know clearly some of the things that must be done."[21]

E.A. Burtt and John Herman Randall, Jr. believed that what must be done was to "search out the reason of things, and not accept them from custom and political authority."[22] This was the religion of the young radicals; their goal was to lead the examined life and transmit their methods of inquiry to the next generation. They could be said to have believed in the Columbia course *Reflective Thinking, a method problem solving.*

JOHN DEWEY AND REFLECTIVE THINKING

John Dewey, at Columbia since February 1, 1905, towered over everyone else in academia in the 1920s. Characterized as a drone in the lecture hall and as an obscure, rambling writer, he was all the same a legend in his own time. Whether other thinkers agreed with Dewey or reacted against him, he was not ignored. His radical pronouncements about history and philosophical truth provided the sparks for discussion and debate, much of it carried on in his absence or in the journals. Dewey's tremendous influence has been documented by J.H. Randall, keeper of Columbia's memories and writer of Columbia's history. Randall's reflection is sentimental, but considering that his own favorite

Professor was not John Dewey at all, but rather, F.J.E. Woodbridge, who was Dean of the Faculties at that time, what Randall says of Dewey probably expresses a true enough opinion of the people who came into Dewey's powerful sphere.

> For Woodbridge, students and colleagues developed admiration; for [Felix] Adler, respect; for Montague, affection. For Dewey it was unashamed love...[It was what he was] that drew men to him--simple, sturdy, unpretentious, quizzical, shrewd, devoted, fearless, genuine--a true Yankee saint.[23]

Dewey had left the University of Chicago in 1904 under a cloud of duress stemming from conflicts around staffing at the experimental school he had set up there. His wife had returned to her job at the school from a trip with Dewey to discover that she no longer held a position at the laboratory school. The complications led to unresolved protest and finally the couple's move to New York City where Dewey took the Columbia appointment.

At Columbia, Dewey put no educational program forward. Secure in himself, with his own developing philosophy and with his growing worldwide fame, he preferred to be left alone. He was unparalleled in the philosophy department. Dean Woodbridge, for whom many apologies for under-recognition have been written, managed the administration. Dewey prided himself on "being a poor man on a committee" and advised his young instructors to follow his lead. With a large department and little co-operation in doing the legwork, it is small wonder that Woodbridge had scarce time for his graduate students or his own writing. From Randall's assessment again:

> It did become Dewey's department; his intellectual leadership rapidly emerged. The university soon looked upon him as the philosophic spokesman of its brilliant social scientists, the great protagonist of creative intelligence applied to human affairs. The Department knew his liberal program of applied knowledge was based on a philosophy that combined the humanistic wisdom of the more critical and scientific strains of the idealistic tradition with the power of the evolutionary genetic analysis of human nature and society.[24]

Dewey's philosophy is a complicated subject and many volumes have been written about it. Suffice it here to give an outline, in general terms, with some implications for history based on the genetic method and reflective thinking since these are the Deweyan fundamentals most directly affecting E.A. Burtt as he wrote *The Metaphysical Foundations*.

Dewey argued that modern science need not be perceived as a break with the past. But science, he thought, had been slow to be adopted by philosophy because "the main efforts of thinkers [have been] inevitably directed to minimizing the shock of change, easing the strains of transition, mediating and reconciling...Men cannot easily throw off their old habits of thinking, and never can throw off all of them at once."[25] Citing Bertrand Russell's popular work, *A Free Man's Worship*, as Burtt did in the opening pages of *The Metaphysical Foundations*, was the code the young radicals used to enter discussions with this point of view.

Dewey preached that new ideas and methods ought not cause any psychic disturbances; they should be made at home in social and moral life. The purpose of the reconstructed philosophy was just that--to mediate between science and social and moral reform.[26] A reconstructed philosophy "can make it easier for mankind to take the right steps in action by making it clear...that intelligence brought to bear upon the observation and understanding of concrete social events and forces, can form ideals...aims, which shall not be either illusions or mere emotional compensations.[27]

Dewey's studies of logic led to his definition of *How We Think*, as a detailed study of the structure of the thought process in problem solving and an account of how the natural, ordinary process can be educated to achieve better results, better solutions to problems with less mental anguish in the long run. Dewey and the other radicals had tremendous faith in human intelligence and in the potential for general education to discipline human minds. Such disciplined thinking would wean them from their natural, unschooled habits and tendencies--especially the compulsive need to resolve psychic tensions as quickly as possible and escape from the very essence of the human condition. In short, the goal of reflective thinking is to learn to accept doubt. Reflective thinking was recommended for everyone, beginning with children.

The reflective method illustrated how to think through the imaginative component in experience and modify it with judgment. Dewey pointed out that because human beings, unlike the animals, remember past experience they live in a world of signs and symbols.[28] No external stimulus, be it emotional experience, perception of the object, etc. can exist in the human mind by itself, alone; there is no literal experience of nature, as the phenomenologists would have it; rather every perception is rich with associations from memory, either from the personal memory of past experience or from the collective impressions organized and categorized by the culture into traditions and social heritage.[29]

"The primary life of memory is emotional rather than intellectual and practical."[30] Experiences, along with their emotional coloring, are organized and generalized into a system of "beliefs" which eventually will lose their connection with experience and require justification in some future historical period.[31] Whether these beliefs are justified rationally by an elaborate metaphysical scheme of logical reference to the Absolute, God, Being as Being or the Universe (the idealist's tendency), or justified by reference to empirical, phenomenal, everyday reality as experienced in practice and described by the methods of modern science (the realists' tendency), any philosophy which seeks justification for belief "has originated not out of intellectual material, but out of social and emotional material."[32]

Dewey argued that the history of philosophy itself is ample demonstration that any philosophy should be understood as a chapter in the development of civilization and culture.[33] The twist that E. A. Burtt was to put on this doctrine was to consider early modern science as philosophy and record its history in *The Metaphysical Foundations* as a demonstration of just another chapter in the story of human intellectual evolution. That he did not take science to be a necessarily true account of objective reality is what makes his thinking ahead of his time. Another goal of the reconstructed philosophy was to integrate the human capacity for judgment with subjective experience. A reconstruction in philosophy would relieve people of having to choose between sense and thought, experience and reason.[34] "Philosophy can not 'solve' the problem of the relation of the ideal and the real. That is the standing problem of life," but it can liberate man from "radically false bias."[35] Logic must be reconstructed, not on mathematical principles or metaphysical principles, but on the human faculty of judgment. "Judgment is the central thing and judgment is not logical at all, but personal and psychological."[36] Dewey did not think that strict logic in human affairs is possible to achieve; instead, judgment alone is possible. The purpose of such a philosophy will be primarily to relieve men's opinions from bias and prejudice. It will lead to a moral reconstruction and a social philosophy dependent on intelligence and judgment.

This was Burtt's point as he systematically pointed out the illogic, which he called the "metaphysics," of Isaac Newton. He was making the argument that all scientific theories, as philosophies, are limited and relative to the culture from which they spring. For him, as for Dewey, it was reflective thinking or meta-philosophy that reveals the nature of reality. And that reality was human reality, only. Unlike the pretensions of science itself, Burtt would make no comment on objective physical reality.

Training the faculty of judgment involves pointing out how we think and what tendencies an undisciplined imagination will favor. In *How We Think,* Dewey gave his own description of "reflective thinking" and his list of certain "Tendencies Needing Constant Regulation."

> 1.) Reflective thought is the "active, persistent, and careful consideration of any belief or supposed form of knowledge in the light of the grounds that support it, and the further conclusions to which it tends.[37]
>
> 2.) Thinking begins with a.) "a state of perplexity, hesitation, doubt; and b.) an act of search or investigation directed toward bringing to light further facts which serve to corroborate or to nullify the suggested belief."[38]

> 3.) "The data at hand can not supply the solution; they can only suggest it." The sources of suggestion are "past experience and prior knowledge." Without prior knowledge or past experience no problem can be solved.[39]

> 4.) "If the suggestion [solution] that occurs is at once accepted, we have uncritical thinking, the minimum of reflection. To turn the thing over in mind, to reflect, means to hunt for additional evidence, for new data, that will develop the suggestion and will either...bear it out, or else make obvious its absurdity and irrelevance....To maintain the state of doubt and to carry on systematic and protracted inquiry [is the goal of reflective thinking]." Reflective thinking requires that initial ideas not be accepted without critical examination just to reduce the time spent in mental uneasiness. Disciplinary training teaches us to suspend conclusions until the problem has been thoroughly thought out.[40]

Dewey described three kinds of unconscious or nearly automatic areas of thought from which the imagination will immediately draw uncritically. With some careful comparison, it can be seen how a modification of Dewey's program for reflective thinking with special emphasis on the "tendencies needing constant regulation" became Burtt's basis for identifying Newton's "metaphysics." Newton, or more correctly, those who arrogated his methods to the British new realism as logical positivism, had not paid enough attention to the "tendencies needing constant regulation." In their rush toward certainty and impatience with doubt they fell into "unconscious metaphysics."

One historian of philosophy has commented, "For Dewey, the term 'pragmatic' meant only 'the rule of referring all thinking, all reflective consideration, to consequences for final meaning and 'test'."[41] Burtt carried this test into the realm of physical science to examine the foundation upon which it rests and demonstrated by reflective thinking that science, even physical science, is a description of the experience of mental pictures derived from nature and subject to the same "tendencies needing constant regulation," Dewey had outlined. By looking at science this way, Burtt leveled science with any other kind of thought. Where Dewey had turned the Baconian method for objective evaluation of phenomena in nature onto thinking in general in order to make all thinking more scientific, Burtt turned the Baconian method from thinking in general back onto the scientific method. The level of prestige accorded to science as a special kind of certainty fell away under this scrutiny and science was driven back onto the same ground from which it had come--theological territory. In the same creative slight of hand, Burtt made metaphysics reappear as a part of the scientific method itself.

Burtt's reforms to metaphysics are by no means entirely derived from Dewey. They are based on Aristotle's metaphysics as inquiry and Kant's analysis of Newton, but the debt to Dewey is a heavy one. He took Dewey's examination of

Francis Bacon's recommendations for clearing away all previous presuppositions in order to evaluate the facts in science objectively, without reference to superstitious belief or previous error, very seriously. Dewey had itemized Bacon's "tendencies needing constant regulation" in the book *How We Think*, a book from which Burtt had taught. According to Dewey, Bacon had recommended avoiding certain "spectral forms which allure the mind into false paths...idols of the tribe, the market-place, the cave or den and the theater." In Dewey's interpretation, now to be applied not just to scientific problems but all thinking, these became understanding erroneous methods (or at least the temptations toward error), which have their roots in human nature generally. These are:

> a.) Those that come from intercourse and language;
> b.) Those that are due to causes peculiar to a specific individual;
> c.) Those that have their sources in the fashion or general current of a period.[42]

In Burtt's historical inquiry into early modern science, these "tendencies" became a thinker's "metaphysics." He described uncritically held metaphysics as "unconscious metaphysics," and identified three main types. The first is what he called the "ideas of an age on ultimate questions," which are uncritically held insofar as they do not run counter to personal interest, in which case the individual might be awakened to criticize them. Burtt says that there has never existed anyone in human history who was free from this particular brand of unconscious metaphysics; but the real philosopher and, therefore, metaphysician, will be constantly on guard against such "idola theatri" influencing his philosophical thinking.

The second sort of unconscious metaphysics is "metaphysics of method." Since every thinker has a method, he "will be under strong and constant temptation to make a metaphysics out of his method, that is, to suppose the universe ultimately [to be] of such a sort that his method must be appropriate and successful."

Finally, Burtt says, "human nature demands metaphysics for its full intellectual satisfaction [since] no great mind can wholly avoid playing with ultimate questions." These sorts of questions are thrust upon every scientist, philosopher or religious individual. Burtt's point was that metaphysics can not be avoided by anyone who thinks at all and therefore, metaphysics must be recognized and accepted as a normal part of the creation of human world-view. But philosophers, in their wisdom, should compensate for it by being self-critical.[43]

Burtt's argument was not about traditional metaphysics; using the word in his bold title was contrived to capitalize on his own idealist frame of mind and the contemporary debate raging about metaphysics exclusion from Russell's philosophy. Burtt offered an imaginative recovery of metaphysics by reforming it as a method of inquiry, "a science which reveals the nature and the place of the intermediate categories between the absolute category reflected by the postulate which all thinkers have to make and categories of the other sciences."[44] This is metaphysics as the study of how we think--based on the absolute, metaphysical reality that we do think. As a method of philosophical inquiry it is an elaborate analysis of the presuppositions at the foundation of any idea set. It has its roots in Dewey's reconstructed philosophy and the genetic method of history as inquiry and reflective thinking-- but no one before or since Burtt has ever used those ideas with as much creativity or skill as Burtt used them.

Dewey did not conceive of his reconstruction of philosophy as a radical break with the past; it was rather a natural "evolution" in human thought, whose time had come. Burtt echoed the same opinion in these words in conclusion to the 1932 edition of *The Metaphysical Foundations.*

> Mayhap we must wait for the complete extinction of theological superstition before these things can be said without misunderstanding....An adequate cosmology will only begin to be written when an adequate philosophy of mind has appeared....an indispensable part of its foundation will be clear historical insight into the antecedents of our present thought-world.[45]

Burtt's point of departure from Dewey was that Dewey, down-to-earth Yankee he was, accepted the mechanical universe as the natural evolution of a world-view suitable for human mechanical intervention and subsequent control over nature. Burtt, on the other hand, deplored the mechanical universe as a perversion of reality, a fragmented picture. Its exposition as transitory made it nonetheless inadequate as any world-view unaware of its own history and limits must be. Burtt wanted to know and explain the history of the present world view; he wanted to use that history to push empirical logico-mathematical science to its limits to redefine the boundaries of scientific reality and, at the same time, to imply the existence of a greater reality beyond, not necessarily physical.

CHAPTER 3

THE REALISTS WHO SHARPENED THE EDGES OF BURTT'S IDEAS

F.J.E. WOODBRIDGE

It was Frederick James Eugene Woodbridge, Dean of the Faculties at Columbia (1912-1929), not John Dewey, who was Burtt's doctoral thesis advisor. Sterling Lamprecht, one of Woodbridge's students, later at Amherst, commented that Woodbridge's teaching "was bound to generate metaphysical doctrines."[1] However, the sorts of metaphysical thinking Woodbridge invited were "analytic in character...concerned with nature and man's inclusion in it. Nothing about this metaphysics was of that transcendental import which current positivism has in mind when it rejects metaphysics."[2] True to the British tradition from which he derived, Woodbridge brought empirical realism to his interest in Naturalism. As a metaphysician Woodbridge was not a skeptical sort. His own work in philosophy was the attempted expansion of Aristotle's categories, prime matter and form, and to substitute for them "structure, behavior, and a natural theology of sequence."[3]

During the 1920s Woodbridge was simultaneously a teacher, supervisor of graduate students, and Dean of the College--obviously a very busy man. Burtt was quiet and Woodbridge was aloof. A good portrait the average student's situation has been drawn by Sidney Hook, in the Columbia graduate program just after Burtt, during the years 1923-1927.

> The primary difficulty with the teaching of philosophy at Columbia when I was a student was that it was insufficiently systematic. Woodbridge was a thinker of deep insight, thoroughly steeped in the history of philosophy, who was convinced that epistemology was a mistake. He was always asking "simple questions," but it required considerable philosophical sophistication to understand the meaning of the questions and preternatural powers of intuition to grasp the "right" answers--which we did by guessing. Dewey at the time was challenging the confusion between cosmic and ethical issues central to the Greek classical tradition in philosophy and the mistaken theories of experience on which the whole of modern philosophy of empiricism rested. Long before Wittgenstein, he denied that there was any philosophical knowledge and dissolved questions like the existence of the external world, the traditional mind-body problem, etc., by showing that on their

> own assumptions they were insoluble or question begging. This approach, as well as that of Woodbridge, would have been stimulating and challenging to students already well trained in the analysis of the traditional problems, but to the miscellany of theological students, social workers, teachers, seekers of wisdom, beauty or social salvation that in those days constituted a considerable part of the classes in philosophy, Dewey and Woodbridge were obscure. Their junior colleagues, some of whom were just as mystified by them, could not dissipate the obscurity.
>
> But to return to the teaching scene. I doubt that the teaching staff got much philosophical stimulation or challenge from those they taught, except in a few small seminars. There was not enough intellectual feedback. Woodbridge enjoyed asking questions that stumped his class but didn't fancy getting questions in return....Everyone except Dewey and Montague seemed to me to be trying to understand why the philosophers of the past said the odd things they did, not whether what was said was true or even formally valid.[4]

Woodbridge continued as Dean of the Graduate Faculties until 1929, when "he resigned as dean...in order to devote all his energies to his primary concern--teaching".[5]

Everyone at Columbia recognized that Woodbridge was a definite second fiddle to Dewey. His role was primarily as well-informed teacher and administrator, who made it possible for the brighter lights at the university to shine. He was an excellent logician, who made his strongest contribution to the training of the philosophy majors; it was generally agreed that Columbia students would not acquire the special mental structure and disciplined habits necessary for rigorous work in philosophy if they did not have exposure to Woodbridge. Coming as he did from Yale, rather than being a Columbia graduate already tutored by Dean Woodbridge, it is possible that Burtt was assigned to Woodbridge's graduate student load in order to "catch him up." As an ardent admirer of Woodbridge, John H. Randall described his general outlook and strengths within the Columbia group.

> Woodbridge was familiar with the philosophical problems posed by the tremendous advances of the sciences; he had lived through the impact of evolutionary and biological thought upon the old certainties, and he was in the midst of the rising currents of experimental psychology. Modern philosophy, he was convinced, the response to the older Newtonian world, ought to be dead; philosophers should be dealing with the newer intellectual issues. But whereas Dewey...was using the newer conceptions of biological, human and social sciences directly to effect the same kind of criticism, Woodbridge preferred the more indirect approach. The two, however, were engaged in pretty much the same critical enterprise; and if Dewey's voice issued the clarion call, and was heard throughout the land, Woodbridge was organizing the solid instruction that would enable the students to understand Dewey's reconstruction and its necessity.

> He would ask a searching question, and woe to you if you did not hit upon the answer--not his answer, but the answer. Then he would comment--and you learned what a disciplined imagination means. He would encourage your suggestion, a twinkle in his eye. And then he would put a simple question--and you realized you were lost; you had failed to think through the consequences of your idea.[6]

Overall, Woodbridge was frustrated by the emphasis the pragmatists placed on epistemology and he was known to complain that philosophy was making states of mind its object. He favored historical discussions over theoretical ones. Woodbridge's conviction that "critical and responsible historical scholarship [was] the prerequisite to original philosophical thinking"[7] was fundamental to his position as a philosopher and teacher, and when Nicholas Murray Butler, who as President of the college had been instrumental in building the philosophy Department, retired, Woodbridge continued Butler's course in the History of Philosophy, teaching it himself until 1925. According to John Randall, Jr.:

> A sound and critical knowledge of the history of philosophical thought, tested by searching written examination, has been since Butler's day an essential requirement for the degree [in philosophy]. Under both Woodbridge and his successor, there was emphasis on the historical context in which problems have arisen and become philosophical--that is, driven men to explore and question their assumptions--and on the historical criticism of those assumptions which they did not themselves criticize.[8]

Woodbridge and Dewey agreed on the value of history, but they differed on what it meant. For Woodbridge careful historical scholarship would show that, "one correct understanding of what has happened does not displace another as truth might displace error, but one supplements and enlarges another. Histories which have gone on before are not undone by those that follow after. They are incorporated into them in a very real way. Historical truth, therefore, when it does not mean simply the correctness of the records of history, is progressive."[9]

In contrast to Woodbridge's scrupulous attention to historical research and detail, Dewey's history was a little rough. History that suited Dewey's purposes could be painted with a broad brush, all the better to be suffused with psychological interpretations for the experience of the present. The most famous Columbia historians, James Harvey Robinson and Charles Beard, were closer to Dewey's relativistic position on history than to Woodbridge's realist one. Between the two of them, they were rapidly rewriting the meaning of history, especially for the working classes. The point was to show how political, economic or social motives underlie historical events and to suggest how received versions of history are rather pathetically naive. Since James Harvey Robinson's *The New History* (1912) and Charles Beard's *The Economic Interpretation of the Constitution* (1913) and their collaboration on a project in European History for New York High

Schools, Columbia University had held the highest reputation for the work and influence of its historians.[10] So it was that F.J.E. Woodbridge represented the tradition of the old guard.

Although Naturalism in American philosophy signifies the belief that the scientific method is adequate to study all phenomena, objective physical phenomena, social phenomena, and human consciousness, itself, one part of the scientific world-view is questioned. That is the dualism between thinking and extension created by a full acceptance of the philosophies of Descartes and Locke. For Naturalists the distinction is artificial but the degree of separation between perception and objective reality varies. On the question of truth, Naturalists can be pragmatists, idealists or realists. The key point for Naturalists is that reality, however it is defined, can be understood without reference to the supernatural. Naturalists absolutely disallow traditional metaphysics. Woodbridge, Dewey, Cohen and Burtt could agree on Woodbridge's statement of the contemporary mind/body problem-- beyond that important distinctions lie. In Woodbridge's opinion:

> Descartes' clear and distinct recognition of the radical difference between extension and thinking and Locke's isolation of the world of human understanding from the world of existing bodies in space both raised the problem of the relation of thought to a world external to thought and different from it.[11]

This kind of dualism sets the thinking mind of man over and against nature. It sets thinking above and separate from the physical world, including the body and experiences of the body, and this creates the problem of perceptions. Perceptions can not be part of the physical world, nor part of the thinking mind and yet must somehow be related to both. According to Woodbridge, this artificial dualism essentially renders philosophy a fruitless, meaningless exercise. Only a rejection of dualism, leading to a unified theory of man in nature can restore an adequate realism to philosophy. Woodbridge defended the real as it is in itself in nature, not in abstraction. He wrote:

> The possibility of natural knowledge is not a problem, but the advancement of learning is. Man names that which is expanded unto his eyes according to its properties...The analyses begin and end in the visible world and my repeated, varied, and perhaps tedious assaults on 'dualism' have been made in order to keep that world from losing its preeminence and finality in the pursuit of knowledge.[12]

The relative degree of emphasis on the mind/body problem as an impediment to knowledge of objective reality separates idealist type naturalists, such as Burtt, from the realist types. Woodbridge's solution was a realist's one; he did not conceive of man in relation to the world, but actually in the world as part and parcel of it. For him "consciousness" meant "cognition" and for him there was some definite object of cognition to be identified. Woodbridge was well aware of the problems posed to explanations of cognition by modern science, but he did not make a contribution to their solution himself. He was content to leave it to others to deal with the "newer" intellectual issues.[13]

In the limited time that Woodbridge had for scholarship he was attempting some kind of synthesis between Aristotle's notion of matter limited by form as actuality and modern science ideas, leaving space for "spirit." Sensitive to the philosophical concerns of his day, Woodbridge sought some kind of synthesis between scientific knowledge and philosophy to save philosophy from being chewed up, lost, and permanently discarded by the new scientific world-view.

Woodbridge and Burtt attacked the dualism inherent in the scientific philosophy of Descartes and Locke from different perspectives. Burtt sought to integrate mind into objective reality by demonstrating that presuppositions, which exist in the mind, affect perception, while Woodbridge was seeking a more adequate definition of the objectively real than dualistic philosophies could provide. In a series of highly respected papers on consciousness, Woodbridge defended the view that true knowledge is a direct perception of the object and not of an idea or image of it. A group of six American philosophers, who called themselves the New Realists, published a joint manifesto, based, in part, on Woodbridge's views and his further suggestions concerning Russell's and Moore's philosophy and aspects of Royce's logic in the *Journal of Philosophy* in 1910.

For his part, Burtt valued science methods and furthermore did not see how its findings could be honestly refuted. But he was a reluctant naturalist and half-hearted realist compared to Woodbridge. He probably did not have much discussion with Woodbridge during the early 1920s on these fine points, even though Woodbridge was, at least technically, his advisor because Woodbridge was too busy. Moreover, it was not Woodbridge's style to tutor anyone or lead them along. He simply asked questions.

Years later, after Woodbridge gave up the time-consuming administrative duties and had returned to full time teaching, he advised another Ph.D. dissertation dealing with Burtt's subject. This was Edward Strong's *Procedures and Metaphysics, A Study in the Philosophy of Mathematical-Physical Science in the Sixteenth and Seventeenth Centuries* (1936). Strong's thesis is much more aligned with Woodbridge's own views than was Burtt's. Under Woodbridge's guidance *The Metaphysical Foundations,* revised 1932 edition, became the point of departure for Strong. Strong attacked Burtt's thesis directly and made the case for the realist point of view in writing the history of science, especially concerning Galileo. Published as a book, Strong's *Procedures and Metaphysics* has been used by formalist historians of science to criticize Burtt's history of science and Burtt's use of the word "metaphysics" to describe Galileo's mathematics as Pythagorean.[14] In

Edward Strong's voice Woodbridge's own realist views are heard, loud and clear, a powerful contrast to Burtt's Deweyan program of reconstruction in philosophy of science.

E.A. Burtt was in virgin territory when he examined the roots of the scientific revolution in the 1920s. By applying the Deweyan psychological implications of history to the historiography of the Scientific Revolution he was creating a revolution of his own. Burtt saw the Scientific Revolution as a unique response to problems presented to thinking man by the physical environment of the earth, combined with a particular cultural overlay.

For his Ph.D. thesis Burtt accepted Woodbridge's realist position that science methods reveal true knowledge about objective reality; but he put a new spin on it. It became a kind of "human reality" which had had a dynamic history. Nonetheless real, because it has been a changing reality, in Burtt's hands the history of science describes physical reality as man can know it--at a particular time in a particular culture. And so, Burtt traced the key ideas of the Scientific Revolution to their philosophical roots revealing an evolution of "world-views," each with its own scientific problems, special language, unconscious metaphysics, implications for man's relationship to the world and certain limitations.

Beginning with the ideal reality of Copernicus, Kepler and Galileo, Burtt demonstrated its aesthetic roots in Neo-platonic metaphysics. He termed the Copernican perspective "mathematico-aesthetic" to separate it from the Ptolemaic-Aristotelian world-view. Descartes' dualism with its emphasis on the material and mechanical led to the mechanical world-view, later exaggerated by Hobbes and Locke. Hobbes insisted on interpreting everything in terms of motions and bodies while Locke's philosophy impressed Hobbes' materialist psychology even further into the collective consciousness. Decline in belief in God followed, owing primarily to Newton, although unintentionally and indirectly, and the "mathematico-mechanical" world-view was in place. Burtt termed the modern view "logico-mathematical" where the world is presupposed to be orderly, rational, and completely accessible to the mind of man through mathematics.

In *The Metaphysical Foundations* Burtt treats the succession of world-views as an actually existing progressive reality. This makes Burtt a realist historian, who like R.G. Collingwood, twenty-five years later, described science and scientific knowledge as a part of a changing historical reality. Loaded onto the Pragmatic view of history this means that scientific knowledge is not true or false in itself, but true or false as an accurate reflection of an historical period only. To the extent that a Deweyan thinker admits absolute reality for science facts, it is as historical facts, insofar as they make the present meaningful.

As a graduate student, Burtt must have had to make his thesis acceptable to Woodbridge, yet Woodbridge's own philosophy is hard to square with the philosophical perspective in *The Metaphysical Foundations*. As an historian, Burtt met Woodbridge's standards, but the conclusions he reached must have stepped heavily on Woodbridge's aversion to idealist epistemology. For example, Burtt asked why someone had not preserved the other half of the Cartesian dichotomy, mind, to be accounted for in terms of purpose and use?[15]

Caught as he was between's influence, his own idealist frame of mind and Woodbridge's empirical realism, where Burtt himself stood on the question of truth in history or truth in the absolute sense does not come through clearly in *The Metaphysical Foundations*. He had adopted Dewey's methods; but had he thought through all of their implications? As an idealist, he was comfortable with his conclusions, that the history of science has been a parade of scientific world-views; but the first edition of *The Metaphysical Foundations* does not spell out any specific philosophical position or any solutions to the many problems it raises. Rather it poses questions in such a way as to imply a pragmatic perspective for the history of science while it demonstrates a powerful method of philosophical inquiry. Seven years later, in the revised edition, Burtt anticipated a new world-view, one that will emerge from a new philosophy of mind "which must provide full satisfaction both for the motives of the behaviorists who wish to make mind material for experimental manipulation and exact measurement, and for the motives of the idealists who wish to see the startling difference between a universe without mind and a universe organized into a living and sensitive unity through mind properly accounted for."[16] If this is accomplished, then, "the new metaphysic, when it arrives, will be [more than] the objectification of the mood of an age, perhaps fitful and temporary, rather [it will be] the reasoned expression of the intellectual insight of all ages."[17]

Some readers of *The Metaphysical Foundations* have assumed that Burtt's use of quotations from Dante's *Divine Comedy* and from Bertrand Russell's *A Free Man's Worship* implies a stronger than average religious conviction to be placed over and above the value of modern science. Given the powerful incentive for philosophers and science writers to avoid poetic language and literary device in order to be taken seriously, Burtt was pretty far out on a limb by today's standards. It is, therefore, important to deal with the literary decision he made in light of his purpose in *The Metaphysical Foundations* and in relation to his own philosophical position at the time. Burtt's emphasis in the 1929 book, *Religion in an Age of Science*, his 1933 support for the first *Humanist Manifesto*, and especially his description of Protestant liberalism--("God is no longer the central fact in religion or the ultimate principle in theology. His place is taken by man's religious experience"--)[18] is powerful evidence that Burtt's own religious persuasion was not reflected by either Dante or Russell. His characterization of himself at this time as an "inchoate idealist" in the autobiographical article for *Philosophy East and West* means nothing whatever in any theological regard. Burtt's idealism, at this time at least, should be identified with his suggested resolution to the mind/body problem, that the distinction between primary and secondary qualities should be somehow abandoned or blurred.

The juxtaposition of Dante and Russell, one out of fashion, the other in, helped Burtt to define the unique historical period he was interested in addressing, the Scientific Revolution, as an epoch of intellectual shift.

> *Just as it was thoroughly natural for medieval thinkers to view nature as subservient to man's knowledge, purpose, and destiny; so now it has become natural to view her as existing and operating in her own self-contained independence, and so far as man's ultimate relation to her is clear at all, to consider his knowledge and purpose somehow produced by her, and his destiny wholly dependent on her.*[19] (Italics Burtt's)

With this, the targets were set for the Deweyan psychological explanations.

To read *The Metaphysical Foundations* correctly is to read the Russellian and medieval world-views as examples only, and not to saddle Burtt with a personal investment in either one of them. Burtt's later scholarly interest in the history of religion should not be read backward into *The Metaphysical Foundations* as has been done by Lorraine Daston.[20] Burtt became interested in religion from a scholar's point of view only after his appointment to a teaching position at Cornell many years later. As he relates it in his autobiographical article:

> One duty I accepted when coming to Cornell was that of teaching a course in the history and comparison of religions. Previously I had been quite ignorant in this field.[21]

It was Morris Cohen's influence that directed Burtt's choice of the literary selections, Dante's *Divine Comedy* and Russell's *A Free Man's Religion*. Burtt needed to set up his problem in order to open the Deweyan-style genetic analysis. The pieces chosen from Dante and Russell are both rich in imagery of the type Dewey had identified in *Reconstruction in Philosophy*. They are full of culturally transmitted signs and symbols.

MORRIS R. COHEN

Morris R. Cohen (1880-1947) was born in Russia in the Jewish ghetto of Minsk. The family immigrated and settled in the lower east side of Manhattan. A small man, proud and smart, he excelled as a mathematics teacher and much later as a philosophy teacher. He became an unofficial instructor at Columbia, never assigned to teach any classes. He was systematically ignored in his applications for a Columbia teaching post. John Herman Randall wrote of him:

> Many of the Department's abler students had been trained under Morris R. Cohen at City College. Through such preliminary education, as well as through the personal affection felt for him and the intellectual influence he exerted on their thought, Cohen might well be reckoned an unofficial member of the Columbia group.[22]

Burtt was already 29 in 1921, 31 by the time he finished his dissertation. Woodbridge was a busy man and very likely Burtt was left substantially on his own to do the best he could. This might be part of the reason why Burtt sought out Morris R. Cohen, as a sort of secondary advisor. It is possible that Cohen thought that his informal advising of Burtt, and possibly others, would help him get his foot in the door at Columbia. Or, possibly, he just enjoyed the mental gymnastics with graduate students. There were none at City College.

One of Cohen's students at City College during the same years that Cohen knew Burtt at Columbia (1919-1923) was Sidney Hook. Like the other boys, Hook was in awe of Cohen's command of knowledge and his adroit verbal gymnastics. Cohen's autobiography, *A Dreamer's Journey* and Hook's profile of Cohen in his autobiography, *Out of Step,* corroborate that Morris R. Cohen was an underdog and a fighter, who used the power of words to suffocate the opposition and convince himself that his opinion was the most valid because he was the most persistent in presenting it. According to Sidney Hook:

> Cohen was an irrepressible talker. He out-talked everyone at every meeting at which an exchange was being carried on and invariably insisted on having the last word. His talk was good, very good, but colleagues had difficulty getting their own words in. Students of course, never minded this, but not all of Cohen's listeners were students.[23]

Cohen was also "the personification of the naysayer"[24] and he could be brutally cruel. In spite of his brilliant mind and his captivating effect on some people, his hypercritical manner with students could be destructive. Hook remembered this incident:

> One day a group of us were standing in animated conversation in the marble corridor of the college after an exciting hour with Cohen. After a while Cohen came along, obviously in fine fettle. Catching sight of us, he came over, turned to one of the students present, and said, "The trouble with you A, is that you can't think." I can still see that student blanch at this sentence of doom pronounced before his peers, and I blush to think that, so great was Cohen's authority with us, we didn't hold this judgment against him but against the hapless victim.
>
> Yes, Cohen unwittingly left scars, and--as I discovered when I ran into this student a decade later--some of the scars did not heal.[25]

Another unofficial presence around Columbia University at the time was Randolph Bourne. Small, crippled from early childhood, and seriously disfigured

Bourne was an energetic, articulate spokesperson for the young radicals and all marginalized Americans, including the waves of immigrant boys studying for college degrees in New York City. In *Youth and Life*, a book of essays, Bourne argued the importance of pre-World War I youth in overhauling American society. He had studied with Dewey at Columbia (M.A., 1913). Bourne thought that "cosmopolitanism," interaction between diverse cultures and subcultures fit in well with the agenda of the radical reformers of the twenties, who viewed the immigrants, especially the most intelligent ones, as "repositories for insights and experiences that can be drawn upon in the interest of a more comprehensive outlook on the world."[26]

> Bourne believed that a process of cultural cross-fertilization was underway, especially among young people: "It is not uncommon for the eager Anglo-Saxon" in college to: find his true friends not among his own race but among the acclimatized German...Austrian...Scandinavian or Italian," for such persons "are oblivious to the repressions of that tight little society in which the Anglo-Saxon so provincially grew up."[27]

This gives some insight into Cohen's popularity and the reasons why his manner was not reproached, but, on the contrary, highly regarded. In spite of the fact that the bright young men were drawn to Cohen, F.J.E. Woodbridge as Dean of Faculties at Columbia could not offer him an appointment. In Sidney Hook's opinion:

> The ambivalent feelings Cohen harbored toward Dewey may have arisen in part from disappointed expectations with respect to Columbia University. He once told me that Dewey, if he had really wanted to, could have arranged for Cohen to join the Department of Philosophy at Columbia. Years later I asked Dewey about this, and it turned out that by the time Dewey had gotten to know Cohen well, he had dropped the reins of control in the department. The person who really barred Cohen from a post at Columbia was F.J.E. Woodbridge, Cohen's first graduate teacher in philosophy--perhaps more for social reasons than for any others. Cohen was a diamond in the rough, unpolished and argumentative.[28]

Although not mentioned by Hook, the possibility must be admitted that it was anti-semitism, plain and simple, that kept Cohen out of a professor's chair at Columbia University. Eventually Cohen went on to teach philosophy at Chicago (1938-1941) where his presence helped to strengthen the intellectual axis between Columbia and Chicago. For a brief time, just before his stroke in 1942, Cohen was at Chicago with Russell and Carnap.

Cohen was famous for his arrogance as much as for his brilliance. It is frequently reported that his arguments dominated any verbal exchange, philosophical or otherwise. It is easy to imagine Burtt framing his argument for his dissertation with references to ideas he and Cohen had discussed in more-or-less one-sided conversations. The pen, in such a case, is mightier than the drowned-out spoken word and there were some genuine points of disagreement between these two.

Cohen's mark was negativity; his criticism was stimulating because it always struck sharply at the weakest aspect of the point in question, even if it offered no improving polish and suggested no alternative solutions. Cohen was suspicious of the genetic method in general, which he criticized as the "genetic fallacy" in *The Meaning of Human History* (1947, published posthumously by Cohen's son, Felix Cohen). He attacked no only the position that scientists can not help but look upon the facts without preconceived ideas, but also argued against the idea that all knowledge is created by man rather than discovered in nature. "Man, however committed a seeker and teller of truth, is "a crooked stick," he would say, "an imperfect being living in a distinctly 'limited' environment."[29] This dictum applied equally to genetic interpretations as to anything else.

Cohen's pessimism sensitized Burtt to the tragic view of history. Cohen saw man as the "tragic agent, destined to participate in evil while achieving limited good."[30] Burtt's Newton is something of a tragic figure--brilliant scientist, anxious theologian, dismal philosopher, made to stand for something he never intended--the eclipse of philosophy as wisdom by natural philosophy as certainty.

A Free Man's Worship was widely read and discussed within the Columbia circle. M.R. Cohen actually preferred the Russellian tragic view while a good many others were distressed enough by it to try to find a philosophic solution to return human beings to a position of power in the universe by helping people deal effectively with cultural change. In fact, this was one purpose Dewey intended for his inquiry doctrines. In the book, *Morris R. Cohen and the Scientific Ideal*, David Hollinger has observed:

> [Cohen] preferred the Faustian struggle against fate, the stoic defiance of inevitable pain and death popularized in this period [the early decades of the nineteenth century] by Bertrand Russell in "A Free Man's Worship."[31]

"The good cheer idealists managed to mix with their sobriety was...a stumbling block for Cohen: great as was his faith in deductive logic, he was socially and psychologically protected against its use to prove that God was in his heaven and all was right with the world."[32] Hollinger says that Cohen "posed as a centrist." There was a favorite way Cohen had of putting it in the years Cohen and Burtt interacted: "Against the blind worship of the dogma of universal and absolute change, it ought to be sufficient to point out that change and constancy are strictly co-relative terms...[and]...The belief that the world consists of all change and no constancy is no better than the belief that all vessels have insides but no outsides."[33]

The predicament for scientific man is that he is compelled to see himself as "the product of causes which had no prevision of the end they were achieving; that his origin, his growth, his hopes and fears, his loves and his beliefs, are but the outcome of accidental collocations of atoms; that no fire, no heroism, no intensity of thought and feeling, can preserve an individual life beyond the grave...etc., etc."[34] Newton is, for Burtt, still the hero of modern science, but also an unwitting stimulus for a demoralizing scientific philosophy, the "mathematico-mechanical" world-view. Burtt made Newton into a pious man, who was a good scientist, but an inadequate philosopher. As a tragic figure in Burtt's hands, "Newton's [own] cherished theology was rapidly peeled off by all the competent hands that could get at him, and...his metaphysical entities and assumptions shorn of their religious setting were left to wander naked and unabashed through the premises of subsequent thought..." The skeptical philosophers, Hume and Kant ("that ruthless destroyer of age-long theistic proofs")[35] helped to popularize Newton's philosophy by incorporating it in their own. Under their "scientific" scrutiny the world became an isolated object, perceived by man, but without any explanation for the purpose of man and his knowing, without any room for the knower among the atoms in motion.

Cohen like his teacher, Woodbridge, believed "everything which [was] wrong with modern philosophy has its roots in this fundamental dualism."[36] Like Woodbridge, Cohen was a realist and he tended toward mathematical realism. The following lines from Cohen's *Preface to Logic* (1944) reflect his point of view.

> The assumption that numbers and mathematical laws are mental is due to the even more widespread notion that only particular sensible entities exist in nature, and that relations, abstractions, or universals cannot have any such objective existence--hence, they are given a shadowy existence in the mind.[37]

Burtt, of course, did not think that numbers were any more real than the phenomena they describe. The contrast between Cohen and Burtt shows up in Burtt's 1932 review and critique of Cohen's *Reason and Nature*. Burtt wants to argue that order and reason in nature are not true as propositions; they depend on the presuppositions of modern science, now deeply ingrained in the modern mind through long-time exposure to the scientific heritage.

> We come then to the hard facts which stand squarely in the way of Professor Cohen's theory. People, ages, races, and groups have differed in the kind and degree of rationality which they have found in their world. Many find much more chaos and uncertainty than the opposite, and it may be surmised that were we not so strongly gripped by our scientific enthusiasm we should agree with them; certainly none are without some experience in which the sense of unpredictability in the cosmos becomes overwhelming. Is it scientific to propose

> an answer to the problem of the rational structure in nature that ignores these circumstances? Would such an answer have any more logical virtue than an answer to the problem of the motion of the sun which ignores the fact that there are different points of view from which the motion of the sun appears differently? And surely such variations depend on the observer as well as on something outside him. Indeed, if we conceive mind as broadly as Professor Cohen does, may we not say definitely that at least some of them depend on the observer's mind? Our author himself admits that people see facts differently according to the different wishes and expectations which guide the seeing. Now wishes and expectations are mental affairs. Do not the above considerations strongly indicate that metaphysical facts are in the same case as the rest, and that therefore some of the variable conditions affecting the rational order of the world are to be located in mind? [38]

> And is it not rather unscientific anyway to put on such haughty airs about the virtues of the perspective from which we orderly minded folk happen to view the world? [39]

> Reason is a substantial factor in the world? Yes, assuredly, and so is unreason. Under certain conditions the one will seem to dominate nature, under different conditions the other. It is important to locate these discriminating conditions, for no explanation of reason in nature will do which forgets them. To be sure, such discriminating conditions themselves will not be found unless we postulate that they, like other objects of intelligent inquiry, reveal an orderly pattern, and for this reason pragmatists will doubtless affirm as vigorously as Professor Cohen that it is highly desirable to discover as much order in the world as we possibly can. Clearly the dependable conditions of unreason must be as rational as the conditions of reason. But it is one thing to postulate rationality in this way as a matter of intelligent method--it is another thing to insist that it exists entirely apart from those conditions which seem to be required for its appearance. Plenty of unreason, too, exists in the world under appropriate conditions, and if we eliminate the conditions how shall we justify ourselves in affirming reason as the substance of nature rather than unreason. [40]

From the earliest days Burtt's work makes numerous asides and allusions to the irrational. For example in the first version of *The Metaphysical Foundations*, the original Ph.D. version, Burtt writes as an aside comment on the gap between theory and practice: "Illogical, of course, but most people do not live by logic."[41] If we make the world totally reasonable by working out a cosmology in the Greek sense as an entirely integrated world view, a scientific world view predicated on perfect orderliness, we will have a metaphysics of method. That is, the world will appear to be what the method of discovery demands it must be in order to allow discovery by that method. For example, if a mathematical method for discovery is used, the world will be perceived as mathematical. If the world is not completely orderly, we are barred from examining the disorderly parts and they cease to exist in the world-view. For Burtt, mathematics is useful in giving a description of the physical world, but it is instrumentally true only; failure to recognize the limit of the mathematical method will limit awareness. Orderliness or reason in nature or in man is allowed as a presupposition of thought only. As uncritically held dogma, it creates a prison of illusion, a world of diminished possibilities.

Cohen held the view that reason coupled with scientific methods was capable of discovering the nature of things as they are in themselves. He thought that science was doing this all the time and moving ever closer to the truth of things, although constant revision was necessary. He did not think that reason could discover the whole truth about the universe; in fact, he held to the notion that the universe as a whole is unknowable. Science can reveal only parts. Cohen might have quipped to Burtt lines he was famous for:

> Our reason may be a pitiful candle light in the dark and boundless seas of being. But we have nothing better and woe to those who willfully try to put it out. [42]

Both Burtt and Cohen held to the idea that philosophical assumptions and philosophical inquiry were inescapable parts of thinking. Surely they discussed the reflective method and examined the presuppositions at the foundation of their own beliefs and in this sense they are the most remarkable thinkers, the most intellectually honest philosophers of all the cast of characters around Columbia in the 1920s. It was Woodbridge who set the standard for Burtt's scholarship in history. It was Randall who gave him support for his speculative turn of mind, but it seems to have been Cohen who provided the critical intellectual yardstick, which launched Burtt into philosophy.

From John Dewey and John H. Randall, Burtt assimilated the reflective method of inquiry; from Dewey's *How We Think* he took his clues about reforming metaphysics along scientific lines; from John H. Randall, Jr. and his father, John H. Randall, Sr., he derived his liberal Protestantism, from James Harvey Robinson and Charles Beard, and through teaching the Columbia course in *Reflective Thinking*, he got the elements of the genetic method of historical analysis; from Woodbridge, Burtt learned the exacting discipline of historical scholarship; from Cohen he got the stimulation of argument and a tragic sense of history. At the end of it all, the whole proved to be more than the sum of its parts; *The Metaphysical Foundations of Modern Physical Science* was and is still today a work of cosmic singularity. Nothing can be seen to have preceded it and no subsequent work, including Burtt's own, has had the same sweeping significance with the possible exception of Thomas Kuhn's *Structure of Scientific Revolutions*. *The Metaphysical Foundations* is a product of its era and yet more. Quietly, over the years Burtt's peculiar philosophy of science with its devotion to identifying the presuppositions of science has penetrated into the crevices of the modern mind and shaped it. The whole electric environment at Columbia University in the early 1920s invigorated Burtt's natural proclivities. There, in the company of like minds, Burtt produced his Ph.D. dissertation, the work that was to become the most influential book of his lifetime.

Although Burtt's thesis was completed in 1923, he was not granted a degree until 1925, due to a technicality in Columbia's rules for Ph.D.s. One hundred copies of the dissertation had to be made available to the school library and Burtt did not have the money for the copying. When *The Metaphysics of Sir Isaac Newton: A Critical and Historical Essay* was published in England by the International Library of Psychology, Philosophy and Scientific Method by the publishing firm, Kegan Paul, Trench, Trubner and Company, as *The Metaphysical Foundations of Modern Physical Science*, a small printing of one hundred copies was done in the U.S. by Harcourt, Brace and Company and this allowed Burtt to get the one hundred copies he needed for the library.[43]

Burtt left Columbia to join the faculty at the prestigious Chicago School of Philosophy at the University of Chicago in 1923. The Chicago school was unique, dynamic and pragmatist through and through. Whatever further influences there were to be on Burtt's development as a philosopher, it was the New York years which liberated him from his father's fundamentalist religious dogma and apparently from all defined traditions. In one form or another the basic question, "What are the metaphysical, methodological and moral implications of modern science?" was to play a role in his thinking there after.

CHAPTER 4

THE METAPHYSICAL FOUNDATIONS OF MODERN PHYSICAL SCIENCE

The Metaphysical Foundations of Modern Physical Science is the first historical account of the scientific revolution.[1] In this book Burtt placed Newton within a context of culture and religion without devaluing his scientific achievement. But, in a surprising way, the Newtonian world-view is thrown into new light. Burtt claims that the mechanical universe with its Cartesian underpinning, that is, the duality of subject and object, and the mathematization of nature, culminating in "laws of nature," is just "the objectification of the mood of an age, fitful and temporary". Furthermore, scientific claims to truth come at the expense of other means of human knowing which we can not afford to lose. Although the book might be read simply as a polemic against logical positivism, which was competing with pragmatism in the U.S. in the 1920s, it should also be considered as a demonstration of a new philosophy of science. In *The Metaphysical Foundations of Modern Physical Science* Burtt is saying that science is a culturally conditioned expression of human experience, just one way of knowing among many, and that scientific truth is a changing truth.

UNDERSTANDING THE METAPHYSICAL FOUNDATIONS OF THOUGHT IS PART OF ACCEPTING CHANGE

There are two lines of philosophical argument in Burtt's thesis presented upon a framework of historical evidence. One is a critique of Newton's *Rules of Reasoning* informed by Burtt's own careful study of Hume. Burtt wanted to challenge the premise of universal order and predictability upon which the scientific method is based. The other is a systematic historical analysis of the scientific world-view. According to Burtt the scientific world view is a perspective on the physical world, but not a true picture. Modern science has become so fully in-grained in us that we no longer think about how we use the legacy from Descartes to interpret our world. It is because of our training that we experience ourselves as the sensate center of consciousness, radically cut off from objective physical reality. We think we experience subjectively the objective real world, which is taken to be the physical world of matter and motion. Burtt argued that we perceive the world more or less subjectively based upon sensory "input," processed by mental function, imagination, and memory. The scale of what we judge (erroneously) as more or less real coincides with what is more or less objective and what is more or less objective coincides with what is more or less quantifiable and mathematically expressible. The more mathematical, the more objective, the more objective, the more real.

In modern-day scientific culture it is common-sense that the world of physical stimulation is separated from the world of ideas. It is part of the scientific world view that we attempt to "know" the world of our experience by dividing it into "primary" qualities--those which inhere in matter, are mathematically quantifiable, and therefore "real" and "secondary" qualities--those which are not quantifiable, less accessible to scientific explanation, and therefore, less "real." Science has expelled the "less real" aspects from meaningful experience. The effect has been to devalue ideas and overvalue mathematical, empirical information.

Burtt undertook his study of the historical development of scientific objectivity to illustrate that it is a only one perspective on perception and not an infallible window on reality. Classical modern science is simply the current operating framework and has no necessary claim to explain reality, whatever that might be, and certainly no claim on future explanations. Later, in 1935, he would argue that it is just a biological adaptation that has evolved and will continue to evolve. In *The Metaphysical Foundations* Burtt made his point and at the same time demonstrated a powerful philosophical method. At the same time, he did the groundwork for his own theory of metaphysics as a study of the categories of mind through which we experience the world as we do.

Important, too, for Burtt's inspiration that the Newtonian world-view is not the last word on physical reality were new developments in physics. Einstein's work was rapidly becoming known. At the time, it must have seemed that the Newtonian world-view was about to collapse just as the medieval world-view had done in the wake of Newton. It is small wonder that the young Columbia philosophers thought of themselves as new humanists ushering in a new renaissance. Burtt's goal was to prepare the world for change.

Einstein's world-view has not become integrated with everyday experience. At the beginning of the twenty-first century, we still live in the man-sized world of Newtonian principles. While we recognize that this is not the real physical world, educated people know only vaguely of that other world, the newer real world of the quantum. So far, it has not affected us much.

Burtt proposed a new definition for metaphysics to facilitate shifts in world-view. Metaphysics would become the unrecognized premises at the bottom of confidently held concepts. Logical and penetrating analysis of the presuppositions of thought, accomplished by reducing interconnected ideas to their simplest relations, was to Burtt the logical extension of the realists' own procedure. It was they, who recommended determining the meaning of a concept by analyzing it into its logical simples fixed by the relations of presuppositions supporting propositions. But Burtt could not accept the realists' "teleological faith" that an "external structure of meaning that is common to all of us" can be revealed in this over-simplified way because "the history of philosophy show[s] that we can [never] transcend our variable biases." Finding out what presupposes what does not get us any nearer to the truth. Burtt saw "that what presupposes what depends on one's point of view and in particular on what he takes the process of explanation to be, while any object is simple if one can offer in terms of it the kind of explanation that one wants."[2]

Burtt's metaphysics, an analysis of underlying cultural or personal presuppositions, was to become a fixture in his thinking. Other philosophers adopted and followed his lead, notably R.G. Collingwood with his own Kant-dependent version, of Burtt's ideas, *Essay on Metaphysics* (1940).[3]

In *The Metaphysical Foundations of Modern Physical Science* Burtt described uncritically-held metaphysics as "unconscious metaphysics," and identified three main types. The first is what he called the "ideas of an age on ultimate questions," which are uncritically held insofar as they do not run counter to personal interest, in which case the individual might be awakened to criticize them. Burtt says that there has never existed anyone in human history who was free from this particular brand of unconscious metaphysics; but the real philosopher and, therefore, metaphysician, will be constantly on guard against such "idola theatri"[4] influencing his philosophical thinking.[5]

The second sort of unconscious metaphysics is "metaphysics of method." Since every thinker has a method, he "will be under strong and constant temptation to make a metaphysics out of his method, that is, to suppose the universe ultimately [to be] of such a sort that his method must be appropriate and successful."[6]

Finally, Burtt says, "human nature demands metaphysics for its full intellectual satisfaction [since] no great mind can wholly avoid playing with ultimate questions."[7] These sorts of questions are thrust upon every scientist, philosopher or religious individual. This is presumably the kind of metaphysics which compels people to use words such as "I wonder." Burtt's point was that metaphysics can not be avoided by anyone who thinks at all. Metaphysics should be recognized and accepted, but philosophers should compensate for it by being self-critical and aware of their own metaphysical leanings.[8]

Burtt wanted to attack the received view of causality using this concept. He repeatedly made the point that the idea of causality has had a dynamic history and he used that history to criticize the position of the positivists. The argument worked on two levels.

First, Burtt's treatment of the history of science suggests that all scientific meaning is in flux. If causality as a philosophical concept has undergone a succession of interpretations, then any current interpretation, even if it is called a scientific one, cannot claim any greater validity than its predecessors or its successors. Burtt suggested that scientific concepts in general are part of a dynamic progression of idea systems.

Second, Burtt showed how a logical extension of Newton's causality had an old metaphysical foundation. If the positivists adopted Newton's logic, they adopted his metaphysics of causality and the positivists' claim for the escape from metaphysics is a false one. According to Burtt, one by-product of Newton's philosophy in early modern science was that God was replaced as first efficient cause of the world by the idea of universal law as formal and only cause of natural phenomena.

Burtt explained how the logical extension of Newtonian philosophy of nature produced a new concept of cause and effect. For example, according to Newton's thinking, it is possible to "reduce the major phenomena of our solar system to mathematical law on principles that do not depend on the presence of the fixed stars, and hence with no reason to suppose that their disappearance would upset our formulations in the least."[9] We actually have no idea what the disappearance of the fixed stars would, in fact, do to our solar system, but in the absence of this knowledge and while the stars are still in place, therefore not testing our formulations, it is still perfectly possible to erect a system of presuppositions and base mathematical formulations upon them. This is possible because we simply narrow the field of cause and effect relationships without worrying too much about unanswerable questions--that is metaphysics. The idea of cause and effect has been transformed by Newton's own "metaphysics of method," which ignores its own metaphysics of excluded hypotheses, so that we understand that certain physical phenomena occur according to laws operating independently of other causes in the universe. The philosophical lesson Burtt draws from this inquiry and example is "even the attempt to escape metaphysics is no sooner put in the form of a proposition than it is seen to involve highly significant metaphysical postulates."[10]

Next comes Burtt's punch at the positivists. "There is an exceedingly subtle and insidious danger in positivism."[11] As demonstrated, metaphysics can not be avoided. Anyone who denies that fact is unaware of his own metaphysics. Such a person holds presuppositions dangerously because they are held uncritically and unconsciously. This is a grave philosophical error because the "unconscious metaphysics" are communicated indirectly to others. Burtt says that they "are propagated by insinuation rather than by direct argument."[12]

> [That metaphysics cannot be avoided] can be illustrated by analyzing any statement you please; suppose we take the central position of positivism itself as an example. This can perhaps be fairly stated in some such form as the following: It is possible to acquire truths about things without presupposing any theory of their ultimate nature; or, more simply, it is possible to have a correct knowledge of the part without knowing the nature of the whole.[13]

By demonstrating how Isaac Newton's most serious students had uncritically adopted his unconscious metaphysics, Burtt's argument implies that logical positivism, insofar as it had adopted the logic of mathematics, could be shown to be as metaphysical as mysticism. The strength of the argument could come from concentrating upon the uncritically held assumptions the positivists had accepted from the dogma of early modern science without examination of their origin within the whole idea set from which science derives and its evolution within the history of ideas. Burtt used the history of science as a mirror for the positivists; the picture he drew reflected the progress of science as a parade of world views built perpetually on shifting sands, one idea system following another and destined to continue the pattern.

Burtt simply refused the notion that modern science is a cumulative enterprise culminating in objective truth about what nature is in itself. He decided that the early modern scientists, renowned for the certainty with which they "delivered to the conquering mind of man" the hard facts of the universe, had a "metaphysical" intellectual foundation. They had based their formulations upon unverifiable "presuppositions," their own "metaphysics," their aesthetic proclivities or their religious beliefs.

> Is there not a flagrant contradiction in such a phrase as 'the metaphysics of Newton?' ...Newton we are told was the first great positivist...He turned his back on metaphysics in favor of a small but growing body of exact knowledge. With his work the era of great speculative systems ended and a new day of exactitude and promise for man's intellectual conquest of nature dawned. How, then, speak of him as a metaphysician?

> To begin with, there is no escape from metaphysics, that is, from the final implications of any proposition or set of propositions. The only way to avoid becoming a metaphysician is to say nothing.[14]

Burtt's historical examination of early modern science challenges scientific method by calling the scientists' belief in an orderly, uniform, and mathematically intelligible universe into question. His intention was not to reaffirm medieval man's favored position in the universe as some have claimed.[15] His purpose was to release modern man from the grip of the scientific world-view in order to make way for what he believed was the way of the future, an even more adequate epistemology. Through a critical analysis of Newton's mechanical science Burtt wanted to expose the logical weaknesses in the newly emerging British analytical philosophy and to undermine the claims the positivists were making to certain knowledge in fields outside of science.[16]

Logical positivism (or new realism) was being put forward by Bertrand Russell and A.N. Whitehead in *Principia Mathematica* (1910-1913). Their ideas were given further dimension by G.E. Moore's essay, *Refutation of Idealism* (1903). Taken together, these two British philosophical works represented a significant change in emphasis, opposed to American Pragmatism, and in complete rejection of existing American and British Idealism.

According to the new realists, only two kinds of statements have meaning; these were logical tautologies and verifiable or falsifiable assertions about empirical facts. The American philosophers and social reformers opposed the British philosophy because it excluded questions concerning the human condition and analysis of social problems. Besides being incomprehensible as philosophy to the Pragmatists, the new realism completely shut out all philosophers with idealist tendencies and interests. The Russellian standards for philosophical truth became coupled with the scientific critical empiricism introduced into physics by Ernst Mach. Both mathematical logic and Mach's empiricism were incorporated into Russell's logic, which then became known as "logical positivism."

Russell and Whitehead had sought to exclude as meaningless most of the traditional philosophical questions because such questions could not be approached by their mathematically inspired methods. For the new realists and the logical positivists, philosophy was no longer defined as the search for wisdom, but set to the business of defining and clarifying meaning for the scientists, who, it was assumed, might otherwise fall into error, not knowing just what it was they were doing without clear direction from the philosophers for verification. In some ways, Burtt can be seen as trying to help accomplish this task by simplifying the presuppositions upon which science depends. His antipathy for the mathematical trends, emerging with the new logic, was definitely there; his hopes to supplant it with something more to his own liking can be teased out from the open-ended conclusion to the revised edition of *The Metaphysical Foundations* (1932).

Burtt's quarrel with Russellian philosophy was not only that it shut out metaphysics; he did not think that the goals for truth in empirical science and philosophy were comparable. In his introduction to his book, *The English Philosophers From Bacon to Mill* (1939), he argued that scientific knowing is "limited [to] fields where the applicability of accepted methods of analysis, measurement, and explanation appears to be assured." He left to philosophy the more speculative areas of the human enterprise.

> At times [the scientist] proffers rather general theories, purporting to account for large masses of fact, but these theories never claim to embrace everything without exception, and they are normally such as established procedures of verification are essentially competent to test. When a theory transcends such limited generality, or when it must devise its own method of verification, we have left the realm of science for that of philosophy.[17]

Burtt has this to say of Isaac Newton, scientist and philosopher:

> In scientific discovery and formulation Newton was a marvelous genius; as a philosopher he was uncritical, sketchy, inconsistent, even second rate.[18]

Burtt's analysis of Newton illustrates his belief that good science need not be aware of its own epistemology, but that good philosophy must be. Not only must an adequate philosophy be aware of its own metaphysics of method and its own presuppositions, it must be useful in working out the presuppositions underlying all kinds of intellectual endeavors, especially science. It was philosophy which could provide awareness of science methodology to science; not vice versa, as the realists proposed. This did not exclude Burtt's admiration for science methods, but rather that he emphasized the need for science to acknowledge its fundamental presuppositions.

Burtt argued that Isaac Newton's science was predicated on the unverifiable presupposition that the universe is dependably ordered. In this sense Newton's science was not science, if science is certain knowledge about the universe. Burtt showed from Newton's *Rules of Reasoning* that Newton considered an orderly and dependable universe to be an assumption. Here Newton had left science for philosophy. The Enlightenment philosophers who adopted Newton's science and built their philosophies around it had wandered into error when they failed to acknowledge the basic presuppositon that Newton had set out.

Burtt made the distinction between philosophy and science that when philosophers ask comprehensive and ultimate questions, they raise them only with careful intellectual responsibility. Burtt wanted to show that while Newton was a brilliant scientist, he was not an especially competent philosopher and that as the Enlightenment thinkers adopted Newton's science, they inappropriately put it forward as philosophy. At the same time, they gave their own ill-conceived philosophy an unauthorized authority by calling it Newton's science.

The juxtaposition of medieval and modern world pictures--Dante's from *The Divine Comedy* and Bertrand Russell's from *A Free Man's Worship* in his introductory chapter, might appear today to be a disparaging rejection of the modern world-view in favor of Dante's ideal vision. But this is not the case. Burtt did his research thoroughly and chose his citations with care.

Burtt's playing off the medieval and modern periods was simply a well-worn pattern in genetic history within the Columbia intellectual milieu and Burtt used it as a kind of high sign to those on the inside of the frequent, on-going discussions about the emergence of a new philosophy. The quotations announce that Burtt is about to make a contribution to that discussion. It does not take much research to note a recurrent pattern of references to Dante's vision in the *Divine Comedy* and Russell's vision in *A Free Man's Worship* (1910) in the history and philosophy published by Americans in the period between the world wars. This was their mutual acknowledgment that civilization had made an unremediable break with the past.

It does not follow that that break was greeted with sadness or horror. On the contrary, the American philosophers welcomed this clean sweep--out with old, in with the new! Even Bertrand Russell believed this and Burtt, as well as anyone likely to be reading his thesis, knew it. For Burtt's purpose, Bertrand Russell became "a representative contemporary philosopher of influence," giving authoritative voice to the clean sweep.

> That man is the product of causes which had no prevision of the end they were achieving; that his origin, his growth, his hopes and fears, his loves and his beliefs, are but the outcome of accidental collocations of atoms; that no fire, no heroism, no intensity of thought and feeling, can preserve an individual life beyond the grave; that all labours of the ages, all the devotion, all the inspirations, all the noonday rightness of human genius, are destined to extinction in the vast death of the solar system, and that the whole temple of Man's achievement must inevitably be buried beneath the debris of a universe in ruins--all these things, if not quite beyond dispute, are yet so nearly certain, that no philosophy which rejects them can hope to stand. Only within the scaffolding of these truths, only on the firm foundation of unyielding despair, can the soul's habitation henceforth be safely built...[19]

Philosophers and intellectuals in 1924 could follow Burtt's selected citations beyond the lines quoted. Following the break in Burtt's quotation, Russell's essay goes on to ask the very same question Burtt asks in *The Metaphysical Foundations*.

> How, in such an alien and inhuman world, can so powerless a creature as Man preserve his aspirations untarnished?[20]

One form of the answer, for Russell, Burtt, and many others was a new philosophy and quasi-religion that eventually came to be known as twentieth century humanism, intended, at first, as a scientific religion, a new faith which rested upon the powers of the human mind. Russell captured the essence of it in the next paragraph, although Burtt chose to omit Russell's hopeful scenario to leave his readers open for his own program. Quoting Russell, as Burtt did not:

> A strange mystery it is that Nature, omnipotent but blind, in the revolutions of her secular scurryings through the abysses of space, has brought forth at last a child, subject still to her power, but gifted with sight, with knowledge of good and evil, with the capacity of judging all the works of his unthinking Mother. In spite of Death, the mark and seal of the parental control, Man is yet free, during his brief years, to examine, to criticize, to know, and in imagination to create. To him alone, in the world with which he is acquainted, this freedom belongs; and in this lies his superiority to the resistless forces that control his outward life.[21]

Although it was essential to the point of Russell's essay, Burtt left this out to set a familiar stage and then present his own work based on the genetic method of history and reflective thinking.

American naturalist philosophers welcomed Russell's sense of the historical moment, they did not appreciate the application of mathematical logic to philosophy. They were critical of logical positivism because it did not admit their favorite method, critical historical study, and did not recognize the importance of experience in all its possible dimensions. The appreciation the American naturalists had for history in philosophy was not Hegelian, it was Darwinian, and therefore, they would argue, it was scientific. They studied how ideas evolve modeled on organic evolution.

In *The Making of the Modern Mind,* John Herman Randall Jr. pointed out the value of comparing the present day to the Middle Ages.

> We have traced those streams of thought and aspiration which have gone to swell the flood of ideas and ideals which dominate the present. There is no final picture, no possible harmonization of tendencies into one great

> symphony of the mind. Such a unity belongs, if anywhere, at the beginning of our story, in the Middle Ages...[22]

Randall, too, quoted extensively from *A Free Man's Worship*, including the very same passages Burtt had used. Randall concludes his sweeping history of ideas with an appeal to the two hallmarks that were to play a crucial a role in American Humanism just on the horizon, "Faith" and "Intelligence." Like Burtt, Randall called on Russell to lend authority to his views. *The Making of the Modern Mind* concludes with quotations from *Why Men Fight*.

> The world has need of a philosophy, or a religion, which will promote life. But in order to promote life it is necessary to value something other than mere life...Wisdom and hope are what the world needs; and though it fights against them, it gives its respect to them in the end.
>
> This is Faith; and whatever storms betide, it will remain a precious heritage of man. But there is one thing of even greater worth than Faith; and that is Thought.
>
> Men fear thought as they fear nothing else on earth--more than ruin, more even than death...Thought looks into the pit of hell and is not afraid. It sees man, a feeble speck, surrounded by unfathomable depths of silence; yet it bears itself proudly, as unmoved as if it were lord of the universe. Thought is great and swift and free, the light of the world, and the chief glory of man.[23]

This, then, sets the tone in which Burtt's *Metaphysical Foundations* must be read.[24] He intended his work to be a reflection of new Hope, new Faith in Reason and new Thought. In the heady days of the 1920s, Hope, Faith, Reason, and Thought were worthy of capitalized initial letters. John H. Randall, Jr. capitalized them in the closing paragraphs of *The Making of the Modern Mind*. In this sense *The Metaphysical Foundations* is very much a period piece from a time when philosophic temperaments saw themselves as thinking warriors, the great heroes of a new age.

Russell reviewed *The Metaphysical Foundations of Modern Physical Science* first for the *Dial*.[25] Russell praised Burtt's history while he discounted his philosophy. The review was not particularly sympathetic and given the fact that Russell's life-style and the amount of money it took to support it caused him to produce large amounts of writing which was, perhaps, not really worthy of his great mind, an outsider might not put too much stock in the review. But taken together with another odd fact, we might better appreciate Bertrand Russell's respect for the work of E.A. Burtt. Russell included a textual reference to Burtt's "remarkable book," *The*

Metaphysical Foundations of Modern Physical Science, in his monumental, *History of Western Philosophy* (1946) and relied on Burtt's interpretation of early modern science to inform his own account in the chapter, *The Rise of Science.*[26]

BURTT'S ANALYSIS OF NEWTON'S MATHEMATICAL METHOD

Burtt made a careful analysis of Newton's work, exposing and criticizing its basic presuppositions. This was not done by a logical analysis alone, characteristic of philosophy, but by logical analysis combined with a new critical, historical analysis of ideas within a particular cultural context. Burtt identified Newton's metaphysics based on Newton's scientific writing only, ignoring the religious material for his discussion, but taking ample mental note of it before formulating his conclusion.

1. The world of phenomena can be reduced to mathematical description and referred to mathematical formulae.

Burtt searched Newton's writings for any clarification or justification of the mathematical method, but he found none. We might expect to find a strong statement of the place of mathematics in philosophical method in [Newton's] Universal Arithmetic, which contains the substance of his lectures at Cambridge in the years 1673-83. In this we are disappointed, his directions on translating problems into the mathematical language being applied only to questions which already obviously involved quantitative relationships.[27]

Searching further, Burtt turned to Newton's Opticks, but his conclusion concerning Newton's failure to explain his method remained unchanged. Newton's eagerness thus to reduce another group of phenomena to mathematical formulae illustrates again the fundamental place of mathematics in his work, but as regards the method by which he accomplished that reduction his statements are too brief to be of much aid.[28]

This was a serious omission, and it became the weak link, opening Newton's philosophy to misinterpretation and misapplication later. In Burtt's view, the mathematical method is just another metaphysics unaware of itself.

2. Mathematical formulae can be verified by observation. By extension of this presupposition, whatever can be known with certainty must be empirically known and at best, should be mathematically verifiable.

Burtt points out that despite Newton's mathematical genius and innovations, Newton's final and absolute criterion for verity was always empirical. He says of him:

> Despite the title of his great work, he had far less confidence in deductive reasoning as applied to physical problems than the average modern scientist....In the Universal Arithmetic, he even intimates that some problems cannot properly be translated into the mathematical language at all....It is not too much to say that for Newton mathematics was solely a method for the solution of problems posed by sensible experience.[29]

Burtt's interpretation of Newton's thinking throughout the whole of the *Principia* was based on the understanding that Newton was, first of all, an empiricist.

> For Newton there was absolutely no a priori certainty, such as Kepler, Galileo, and pre-eminently Descartes believed in, that the world is through and through mathematical, still less that its secrets can be fully unlocked by the mathematical methods already perfected. The world is what it is; so far as exact mathematical laws can be discovered in it, well and good; so far as not, we must expand our mathematics or resign ourselves to some other less certain method.[30]

It was never Newton's intention that the mathematical methods he employed should be applied to anything other than physical problems. They were essentially a helpful tool in the reduction of physical phenomena, and he never considered the mathematical formulae any more real than the phenomena themselves. This, of course, was a heresy against the mathematical metaphysics of Galileo and Descartes. The error in philosophical method committed by early twentieth-century logical positivists was that they took over Newton's mathematical methods in their metaphysical guise, an aspect which had been substantially abandoned by Newton. The positivists united the metaphysics of Renaissance Pythagorean mathematics with Newton's absolute empirical verities and formed an "unconscious metaphysics" for the mathematical method. This opinion became the basis of a new metaphysics. The whole intellectual structure was further influenced by Kant, whose keen analysis stripped Newton of his theological biases. Thus, the opinion that mathematical methods assure certain truth passed into the general current of intellectual life in Europe. It was taken for granted because "it insinuated without defensive argument" that it was based upon "unquestioned certainty from the clear demonstrability of the mechanical or optical theorems [of Newton] to which it was attached."[31]

3. Hypotheses have no place in Philosophy.

This presupposition is the basis of Burtt's criticism of Newton and key to exposing the philosophical errors made later by those who became Newton's most ardent students. To illustrate Newton's position on hypotheses, Burtt quoted from the end of *The Principia*:

> "Whatever is not deduced from the phenomena is to be called an hypothesis; and hypotheses, whether metaphysical or physical, whether of occult qualities or mechanical, have no place in experimental philosophy. In this philosophy particular propositions are inferred from the phenomena, and afterwards rendered general by induction. Thus it was that the impenetrability, the mobility, and the impulsive force of bodies, and the laws for motion and of gravitation, were discovered."[32]

Burtt's contention is that no one, including Newton, can escape hypotheses, which he wants to call metaphysics. Burtt asserts that the very nature of thinking, itself, demands presuppositions from which it goes forward. In this Burtt is subjecting Newton to Dewey's reconstruction in philosophy. Burtt's innovation and modification to Dewey's thinking was to analyze basic presuppositions and refer to them as a cultural or individual "metaphysics," either unconsciously, or at least uncritically, held. As a graduate student at Columbia, teaching "Reflective Thinking", a derivative of Dewey's Pragmatism, he absorbed the notion that ideas are "instruments" of thought. The evidence does not support the contention that Burtt considered himself one of Dewey's followers at this early stage, in fact, he thought of himself as advancing a philosophy of his own making, opposed to Dewey's naturalism. But the conclusion that he had been deeply affected by Dewey's logic can not be escaped.

4. Universal laws can be derived by extending the mathematical formulae to cover all identical phenomena.

This presupposition was based upon the legacy Newton had inherited from his ancestors, his "cultural metaphysics" as a Christian and as an ideological descendent of Copernicus, Kepler, Galileo, Barrow and Descartes. Newton could never have thought it possible to deduce universal principles of motion from phenomena if he had not accepted on faith that nature has a simple, uniform mathematical order. According to Burtt, this notion actually arose from a religious and cultural presupposition with an historical basis in the metaphysics of monotheism and the many metaphysical prejudices toward mathematics that had been carried over from the Renaissance revival of Pythagoreanism by Copernicus, Galileo and Kepler.

Burtt determined that there was a contradiction between the principle of uniformity in nature and Newton's own contention that he had freed science from hypotheses, and he turned to Newton's "Rules of Reasoning in Philosophy" for some clarification. The first rule is the rule of simplicity: "We are to admit no more causes of natural things than such as are both true and sufficient to explain their appearances." The second rule is: "To the same natural effects we must, as far as possible, assign the same causes," or taken in a more advanced application: "Where different events are expressed by the same equations, they must be regarded as produced by the same forces." So far so good, but in the third rule Burtt finds the hint of another hypothesis that goes far beyond the strict empiricism Newton thought he was describing. The third rule is:

> *"The qualities of bodies, which admit neither intention nor remission of degrees, and which are found to belong to all bodies within the reach of our experiments, are to be esteemed the universal qualities of all bodies whatsoever."*

Burtt asks on what basis can it possibly be legitimate to generalize "ad infinitum" from immediate, particular individual experience? Newton had justified the third rule by combining the experimental method with the principle of uniformity in nature; but Burtt had shown that this principle was a metaphysical one, predicated upon presuppositions from Pythagoreanism about mathematics and from Christianity about God. According to Newton's own methods, unless the principle of uniformity in nature depends on observation, it is a hypothesis. Burtt finds that Newton's third rule must be interpreted as falling under his fourth rule. The fourth rule is: "In experimental philosophy we are to look upon propositions collected by general induction from phenomena as accurate or very nearly true, notwithstanding any contrary hypotheses that may be imagined, till such time as other phenomena occur, by which they may either be made more accurate, or liable to exceptions. This rule we must follow that the argument of induction may not be evaded by hypotheses."[33]

This can only mean that we have no metaphysical guarantee whatsoever against there appearing exceptions to even the most confidently adopted principles. Empiricism is the only and ultimate criterion for verification, according to Newton. The extension of this thinking is important for an analysis of scientific realism. It seems that certain isolated phenomena can be found to correspond to mathematical formulae by experiment without reference to a simple, uniform mathematical order. Since the existence of a universal mathematical order is an idea which cannot be verified by observation, this means that modern science, since Newton, has accepted one orderly universe governed by universal laws on faith, which is to say, hypothetically, and therefore, subject to possible empirical refutation.

The double irony, which Burtt played upon in Newton's title, *The Mathematical Principles of Natural Philosophy*, was that although Newton thought he had escaped hypotheses, his entire method of deducing universal principles from phenomena was based on the theology of monotheism and the metaphysical mathematics of the Renaissance. Furthermore, the same metaphysical thinking about mathematics affected the British analytic philosophers, but they, unlike Newton, who did not deny theology, failed to advance any other metaphysical system. They, unlike Newton, were not careful about setting the limits of their methods. By trying to escape metaphysics, positivism fell into a metaphysical trap and philosophical error. Burtt maintained that metaphysical thinking is fundamental to the very nature of thinking,[34] and the attempts of positivism to exclude it entirely made some aspects of that philosophy "pitiful, inadequate, or even fantastic."[35]

Burtt's radical skepticism is owing in part to his close reading of Hume and in part to his deep sympathy for Hume's attack on the design argument.[36] I am grateful for this insight to the work of James E. Force, who by relying heavily on Burtt's interpretation of Newton Rule's of Reasoning to reach his own conclusions regarding the relationship between Newton and Hume, inadvertently shed light on the inspiration Burtt drew from Hume. Force, of course, had no idea of the intimate connection between Burtt and Hume.

For Hume, the Design argument rests on Newtonian causality: "like effects prove like causes." Following this scientific premise, the mind of the designing Deity is revealed in His creation, at the same time proving His existence through the material facts of existence. For Hume, according to Force, the design argument is predicated on the unsupported assumption that the works of man, as scientific understanding, and the works of God, as nature, are clearly analogous. Hume sets out to destroy that argument by showing that God and nature are not analogous if Newton's rules are accepted. The strength upon which Hume's argument is built is that Newton's rules were unequivocally accepted in the eighteenth century. According to Force's reading of Hume, Hume uses Newton's Rules of Reasoning in just the way Burtt used them to argue that a conclusion can not be transferred, unqualified and untested, from any part to the whole. There is no empirical proof that nature is uniform with respect to causes, we forever await the instance where the expected effect does not follow the cause.

Force explained that Hume thought "it perfectly acceptable to reason from the part to the whole so long as either the scientist or the design theorist recognizes that this principle ought to have the ontological status given it by

Newtonian design theorists, i.e., a feigned *a priori* hypothesis about the way the world really is. All design theorists, in sum, read the Second Rule of Reasoning without the crucial qualification of Rule Four."[37] By pressing this case, Force makes a tortured reading of Hume as he borrows heavily from Burtt. Burtt had made this argument central to his thesis in *The Metaphysical Foundations.* Burtt took Hume's argument against the existence of God, rearranged it and reframed it so that the presupposition of the existence of God became replaced with the fundamental presupposition of early modern science, the order and intelligibility of the universe. Burtt turned this around. Just as Hume had thrown doubt on the reality of God by turning Newton on Him, Burtt undermined scientific reality by turning Newton on Newton's philosophically irresponsible followers and then onto modern science, itself.[38]

THE HISTORY OF EARLY MODERN SCIENCE AND THE SIGNIFICANCE OF THE MATHEMATIZATION OF NATURE

It was Burtt's contention that the mathematization of nature which made it possible to predict events has held us captive because it gives us a sense of control over nature.

To illustrate that mathematical treatment of nature was instrumental, in the Deweyan sense for Newton but became an odd kind of unsupported number mysticism for the new realists, Burtt placed Newton's mathematical science in an historical context.

First: The certitude modern science grants to mathematics has its historical roots in Renaissance Pythagoreanism and in Christian monotheism. The Aristotelian first efficient cause of phenomena, that is God, was still in place for Newton, but the formal cause of phenomena became transformed by the Renaissance metaphysics of mathematics. It is Kepler's underlying mathematical harmony of the universe, discovered in observable phenomena, which explains why things are as they are. The laws of the universe as causes of phenomena are conceived upon the idea of the one orderly universe of nature, created by the one, intelligent and perfect Being. Therefore, the idea of laws of nature as causes of phenomena is still a metaphysical idea.

Second: Newton held his theology separate from his physics in the sense that for him mathematics was only a tool to be used in the discovery of the laws of nature. Newton did absolutely presuppose one simple, uniform, mathematical universal order which could be revealed to human understanding through the laws of nature; but he could be read, as Burtt chose to read him, holding out this hypothesis as a necessary one and subject to radical revision if it should prove to be contradicted by experiment.

Third: Because Newton failed to justify completely his mathematical methods, he invited the philosophy of Immanuel Kant and David Hume, whose philosophical brilliance disentangled God and metaphysics. Whereas their skeptical critique of certain knowledge, as attainable through rational processes, left the quest for certainty decidedly shaken, no one attacked the mathematical certainties of Newton. As Burtt put it: "The claim of absolute and irrefutable demonstration in Newton's name had swept over Europe, and almost everybody had succumbed to its authoritative sway."[39]

Fourth: The neo-realists of the early twentieth century abandoned metaphysics as a method of philosophy in favor of the methods of mathematics; but these methods are metaphysical in themselves. Because the mathematics became fused with the empirical verities of Newton's physics, the neo-realists became confused about the metaphysics of their method. Newton applied his methods of mathematical formulation concerning bodies, motions and forces to empirical realities, not to philosophical ones; and even in accomplishing this much, he did not really banish metaphysical presuppositions altogether.

Pragmatist readers were Burtt's initial audience. Part of the reason Burtt argued his thesis as he did was to restore philosophy to its own essentials, which the Deweyians believed to be self-criticism; part of the reason was to illustrate the genetic method of history which evolved into the history of ideas after Lovejoy. But the primary intention was to initiate a new direction in philosophy of science, and search for an "adequate philosophy of mind" to support it.

H. Floris Cohen has made the accurate observation that Burtt's aim "was not to write a historical overview of scientific ideas in the course of the seventeenth century Revolution in Science, such as was to be done later by Dijksterhuis, Butterfield, Hall and others." It is easy to forget this because Burtt's book covers so much historical ground, it has the appearance of a sweeping overview of history and this has been misleading to some of Burtt's critics. Cohen rightly contends that Burtt "wished to focus on the philosophical underpinnings and implications of these scientific ideas." For those readers who do not admit any philosophical undercurrents for the achievements for Copernicus, Kepler, Galileo, or Newton, and instead, want to see these figures as science heroes in the positivist mold, Burtt's book "is necessarily devoid of substance."[40] The point Burtt wanted to make was that no scientist, insofar as he is a thinker can escape philosophy, and that the philosophical implications of early modern science were adopted wholesale into the modern world view. This is what makes the book important as a history of ideas; it was not intended as a history of scientific achievement. Burtt was a careful reader of sources. His insight and intuition were keen. As he often observed himself, he was a pioneer. Burtt blazed trails that others would follow.

Burtt became a point of departure for historians with a strong positivist bent and for those who wanted to defend mathematical realism in science. One of these was E. W. Strong, a graduate student at Columbia, advised by F.J.E. Woodbridge, who had also been Burtt's advisor. Strong and some others have been guilty of not always giving Burtt the credit he deserves for his ground-breaking work as they filled out his thumb-nail sketches of the chief figures in his history of early modern science.

Burtt made Newton the centerpiece of his history of science in order to serve a particular purpose. Early modern scientists could not have revolutionized our understanding of the world nor fathered the scientific world-view had they not made the world quantifiable. By pointing out the origins of mathematical realism as having been based on earlier metaphysical ideas, he could undercut the persuasiveness of the mathematical conception of the world. Burtt checked-out the presuppositions of mathematical realism only to find them tied to theological presuppositions. The historical underpinnings of his argument were, to borrow a term from Dewey, "instrumental" in their intention. Burtt's anti-formalist, anti-intellectual agenda was in keeping with the notion that philosophers had a duty to the democratic, generally well-educated American society. Scientific knowledge and scientific methods would soon be changing the world, increasing material comfort, placing the world under human control, shoring up human power while undermining the human spirit. As a philosopher with a conscience Burtt intended to put science in its place as one way of knowing among others. The philosophical program which grew out of *The Metaphysical Foundations* centered on explaining how it is that we think scientifically with the intention of bringing how we think at all to terms with the special brand of thought we call scientific. If a philosophy of mind which explains how we think scientifically were developed and then applied to other kinds of thinking with satisfactory results, the spiritual side of human experience might be saved by the very thing which seemed to threaten it. The material and the spiritual would come together to serve human needs in a new way, under the control of human reason, guided and explained by philosophers. In this way the new history and philosophy of science could guide right thinking as part of the educational program at Columbia University.

This is the point of *The Metaphysical Foundations of Modern Physical Science.* All the rest, rich though it is as history, is ancillary to Burtt's intention in writing his doctoral thesis. Perhaps his explains why he did not pursue the history of science further, but instead, put his energy into a scientific theory of cognition. Although Burtt's thumb-nail sketches of the heroes of early modern science are firmly grounded in the available source material and convincingly drawn, they were intended as instruments of instruction, designed to serve his developing philosophy of mind. His sketches are interpretative history, speculative, and daring. It is not surprising that they came into conflict with the Whig history of science of the 1930s, 1940s and 1950s.

The fact that *The Metaphysical Foundations* has became a classic and remains current is owing to the fact that the questions it raised have yet to be answered. It expresses faith in human reason and the human ability for analysis of human experience in full recognition of the limits of the human condition. Burtt did not deny God, nor did he affirm God. His point was to affirm human beings in their search for the spirit. Burtt thought that one phase of that search was for mankind to pass through the scientific world-view. Like the other young radicals, Burtt believed the journey would be difficult and damaging to the spirit. His task, as he saw it, was to clear the path and soften the trip for those who could follow him.

Over the next few years, Burtt began to work out a theory of categories of perception and cognition, based on our experience of the world from a scientific point of view. He hoped that this working explanation of how human beings experience the material world would be applicable to how human beings experience the spirit beyond the material world and ultimately make a place in consciousness for a new kind of awareness of the spirit and our need for a spiritual life.

CHAPTER 5

THE CHICAGO SCHOOL OF PRAGMATISM

At the turn of the century, philosophy at the University of Chicago was something unique in the United States, an actual school of philosophy in its own right, according to William James. The special brand of Chicago philosophy was American Pragmatism, which shone brightly during the progressive era of the 1910s and 1920s and then faded into obscurity during the 1930s and 1940s. Part of Pragmatism's success was due to the fact that it captured the general anxiety of the age by confronting the tension between science and religion. Pragmatism attempted a kind of resolution. The scientific method was championed and elevated into a concrete ideal to be applied to societal problems through new applications in education, psychology, philosophy, and history. An integrated approach was advanced with the promise that a holistic view would lend new dimensions to what had been the domain of religion alone, such as ethical questions and prescriptions for responsible living.

Part of the explanation for the decline of American Pragmatism during the 1930s, and the rise of analytic philosophy in the 1930s and 1940s was the disintegration of the Philosophy Department at the University of Chicago under pressures brought to bear by President Robert Maynard Hutchins. Hutchins championed a bold revival of the classics at the expense of the modern, forward-looking, and science-based program put forward by the Chicago Pragmatic Philosophers.

Pragmatism did not hold up the past as a model for the present. For those intellectuals who rallied around Pragmatism, uncertainty and a universe of change were the only givens. To this they brought their faith in the integrity and durability of inquiry and, at the same time, their understanding of the tentativeness, fallibility, and incompleteness of knowledge. The pragmatists adopted the doctrine that knowledge was transient and as such just another aspect of the universe of change. In such a world the attitude of inquiry itself offered the most reliable and stable guide. The spirit of science, its method and openness toward the human experience, were welcomed as the new foundation for an advancing culture.

The scientific habit of mind, a favorite phrase summing up Dewey's perspective, was closely allied with skepticism about knowledge, both old and new. From the Pragmatists' point of view the process of inquiry was more important than the facts discovered by investigation because the inquiry process was assumed to have been standardized by science, and thereby, made primary and enduring, while the facts of knowledge themselves were only temporal and subject to change.

The Pragmatists' theory of truth included the notion that truth is plural, and that it is more of a condition that happens to conscious beings experiencing the world than some absolute, rarefied entity standing above human apprehension. In this, the Pragmatists were placing empirical, but perhaps individual, or at least local, experience over and above classical notions of Truth as universal, the kind of truth that implies a special alliance with God and the Absolute. The Pragmatists rejected what they saw as stagnant truth, inherent in ideas already established, and sought truth instead in the process of the emergence of an idea in history as a cultural possession of the collective of human beings who produced it. For the Pragmatists the world was both socially and physically constructed. Moreover, they asserted that scientific inquiry, inter-married with democratic principles, could sustain culture and, over time, change the world. In Dewey's instrumentalism, for example, ideas become true when they function effectively in open inquiry. A measure of truth for ideas is that they have the power to transform the environment.

Here was a philosophy that harmonized with E.A. Burtt's perspective in *The Metaphysical Foundations of Modern Physical Science*. John Dewey had guided the development of Pragmatism at Chicago and it was Dewey at Columbia since 1905 whose special influence reached Burtt through his teaching assignments there. The University of Chicago had been founded by the American Baptist Education Society to be a national university for Baptists, although the Society exercised no financial control over the university, once it was established. Burtt's Deweyan-style Pragmatism and his Baptist heritage fitted nicely into an academic post at Chicago and he became an instructor there in 1923.

Although the school was founded by a conservative religious organization, Chicago rapidly became a hub of scientific activity as the spirit of Pragmatism spread throughout the school. In addition to Dewey's educational and social theories, Chicago Pragmatism was fortified by the religious psychology of Edward Scribner Ames, who had received the first Ph.D. granted by the university in the Department of Philosophy in 1895, and the critical approach to the study of the Bible fostered by William Rainey Harper, Chicago's first University President. These two combined to produce one of the main trends of the new liberal Protestantism, which had attracted Burtt while he was studying at Union Theological Seminary and associating with the Randalls of New York City.

The University of Chicago officially opened its doors for the first time in 1892, the same year Edwin Arthur Burtt was born. Large sums of money had been poured into the project ($35,000,000 from John D. Rockefeller alone, by 1910). President Harper, encouraged by a dedicated Board of Trustees, searched out and hired 120 of the country's brightest scholars to teach in the graduate school and various professional schools. The country's luminaries were lured to Chicago with high salaries and something more. Harper "promised them more freedom in developing their departments and their researches than they could find in older institutions."[1] T.V. Smith, who went to Chicago in 1921 as a graduate student and stayed on to retire from the faculty, commented on Harper in his autobiography, *A Non-Existent Man*:

> William Rainey Harper had founded the University on what would now appear a shoestring of fact, but he maintained it upon an immensity of faith. He pioneered in many concrete ways but most outstandingly in creating morale for cooperative research. And except for the interregnum of President Hutchins, this creative harmony has characterized the University through-out its existence.[2]

Much of the credit for the Chicago outlook belonged to John Dewey, who had arrived there from the University of Michigan in 1894 to become head professor of philosophy. Dewey joined James Hayden Tufts, who had been head of the Philosophy Department from its earliest days. Tufts had been associated with Dewey at Michigan State University. It was Tufts who had recommended Dewey to Chicago in the first place. After his arrival Dewey brought in George Herbert Mead, also from Michigan, as assistant professor and James Rowland Angell, from the University of Minnesota. Angell was named assistant professor of psychology. These men formed the core of the Chicago School, and over the next decade they trained a group of disciples. Among these was Edward Scribner Ames, who did his graduate work with Tufts and Mead, and Addison Webster Moore, who came from Cornell and studied with Dewey. Eventually Moore took over the teaching of Dewey's logic courses completely when Dewey left Chicago for Columbia in 1905.

A remarkable thing about the Chicago School, besides the fact that it was philosophically homogeneous, was that it had succeeded in bringing ideas out of the academic setting and into the world where ordinary people studied or worshipped. But the School was always more interested in science than religion. It was Ames who brought the liberal religious ideas associated with Pragmatism out of his classroom and into church. In 1900, he became minister of the Hyde Park Church of the Disciples of Christ, which stood adjacent to the school campus. Meanwhile Dewey's thinking penetrated into generations of school children. There was an experimental elementary school set up as a laboratory on learning where scholars from all over the university came to teach and observe. There they became familiar with the educational theories of Pragmatism. Mead, Tufts, and Angell were directly involved with the Laboratory School, which was under John Dewey's direction.

In 1905, the core group changed its configuration, but Chicago Pragmatism only grew stronger in conviction and better at defending itself as unique in the world. Angell moved away from the center of the group when a separate department of psychology was set up. Dewey left for Columbia University in the same year, following a disagreement with President Harper over staffing at the laboratory school. Ames was drifting away and devoting most of his time to his Hyde Park Church. During the changing of the guard Moore, more than anyone else, became the spokesperson for the Chicago Philosophy. He made a considerable reputation based upon his unique abilities in clarification and distinction which he applied to Dewey's thinking. By dealing with Dewey's instrumental logic Moore became "after Mr. Dewey the most important and most authoritative member of the so-called Chicago School," according to George Herbert Mead.[3]

Mead, Moore, Ames and Tufts all had long careers at Chicago and all retired from the school. In 1929, Moore retired and died unexpectedly the following year. Tufts retired as chairman of the department in 1930 under duress, but continued to teach half-time. Mead was eligible to retire in 1931, but did not plan to do so, until events forced him into it. He died shortly thereafter. Tufts retired completely after the Spring session in 1932 following the so-called crisis and collapse of the Chicago School of Philosophy.

E. A. Burtt was a key figure in the Chicago school from 1923 until December 1931, when he resigned, along with Mead, Tufts, Arthur E. Murphy and Everett Hall. Certain extant documents suggest that while Burtt was not in any administrative capacity in the Chicago Philosophy Department he acted as spokesperson for the disgruntled group during their final days. He was certainly the only young member of the Pragmatist group in the early years of a promising career. Therefore, Burtt had, by far, the most to lose when the department fell apart.

T.V. Smith had joined the faculty of the Chicago group as an assistant dean in 1923. He had been hired onto the faculty along with Burtt, who came in at the bottom and worked his way up the ladder over the next seven years. By the time Burtt arrived Smith had already been at the University for two years as a graduate student. According to his own account, Smith made a meteoric rise. A year after he and Burtt joined the faculty, Smith as Dean, Burtt as instructor, Smith was made assistant professor, and two years after that, a full professor. There is a hint of some tension between Smith and

Burtt, although Smith makes no mention it directly; he does not even mention Burtt anywhere by name. He does say this:

> I began to get all the "breaks," until, [University President, Max Mason] ... called me into his office and broke the news that I was to be jumped at once to a full professorship, with salary doubled....Mason terminated the happy interview by remarking that I had "guts" and that he wanted men like that at the University of Chicago. So whatever criticism it brought to him, and whatever envy it subjected me to, the promotion was ordered and it stood.[4]

Burtt might have been envious, or at least, privately injured. But Burtt does not mention T.V. Smith anywhere, just as Smith never mentions Burtt.

Smith was the only member of the Philosophy faculty who did not resign in 1931. Darnell Rucker explains Smith's position, staying on in the wake of mass exodus, this way:

> T.V. Smith did not join the exodus since he had no close ties with the older members of the staff. In fact, by his own account he had been retained on the faculty at the request of members of the university administration rather than of the department.[5]

Arthur E. Murphy was hired into the Philosophy Department in 1927. Although he was not as committed to the major ideas of the Chicago School of Pragmatism as the others, he, too, resigned when the Department broke up in 1931. Everett Hall was hired into the Philosophy Department in 1929; he, too, resigned during the crisis of 1931. When Murphy and Hall were added to the faculty, they diluted what had been monolithic pragmatism at Chicago. But by that time, the school was very near the end of its era anyway.

Since Smith went to Chicago in 1921 as a graduate student and Burtt came along two years later, the department Smith describes can stand for the one Burtt joined. Smith says he transferred from the study of literature to philosophy, where, at Chicago, "one could more nearly follow the life of imagination without scholastic debiting."

> And what a department Philosophy was in those days! It had not the scholarship of a Yale, and certainly not the variety of a Harvard, where James, Royce, Santayana, and Munsterbery, not to mention Palmer, had held forth in friendly profusion. Chicago had enough scholarship, but had little ideological variety....
>
> The four senior professors of philosophy at Chicago were all Pragmatists....
>
> Of the four professors (all tall and personally imposing), Edward Scribner Ames was the most agreeable personally, and George Herbert Mead the most stimulating intellectually. In between was Addison Moore, a clever

dialectician and an ardent devotee of this philosophic cause, and James Hayden Tufts, personally austere but punctilious in mastering what earlier men had thought and said.[6]

Mead has been characterized as "a gentleman of the old school, shy, reticent, good-natured, philanthropic; and, above all, always on the grow."[7] His influence was powerful, both on Smith and on Burtt. Burtt kept Mead's photograph on his study wall all his life. Quiet, shy and unassuming is a character description befitting George H. Mead, John Dewey, and also E.A. Burtt. At the same time, they must have been firm, determined, and dedicated to their special cause. It is hard to imagine quiet, unassuming men accomplishing what they did without monumental inspiration and a sense that their work was destined to change the world. This is surely a theme in Burtt's career and it helps to explain just why President Robert Maynard Hutchins' proposals to change the direction of Chicago Philosophy caused such an uproar.

The most complete account of what happened at the University of Chicago precipitating the crisis of 1931 is Gary Cook's Epilogue to *George Herbert Mead, the Making of a Social Pragmatist* (1993). Cook notes that he sent his account of the episode to Burtt to doublecheck it against Burtt's memory in 1987. Burtt concurred with it entirely. Cook had also sent the draft to Mortimer Adler. "Adler did not respond, but Burtt replied as follows: 'I have read with care the chapter you sent me. My memory does not include some of the episodes you mention, but it tells me nothing incompatible with what you say about them.'"[8] Cook's account closely follows a statement Burtt had prepared on behalf of Professors Mead, Murphy and himself. This document, titled *A Statement from the Department of Philosophy*, closes with the resignations of all three professors.

The trouble seems to have begun even before Robert Hutchins assumed his duties as President of the University in November 1929, perhaps as early as May of that year. Mortimer J. Adler was at that time a young instructor in the department of psychology at Columbia, where he had done graduate work in both psychology and philosophy. Apparently he was judged not to have had the mettle for philosophy, and one gets the impression there was more than just a little resentment on his part for having been denied a place in the rapidly growing department at Columbia. Columbia Philosophy, in John H. Randall, Jr.'s recollection, was adding new members with utmost care and consideration for their merit and promise in order to compete with Chicago in prominence. F.J.E. Woodbridge made the final selections into the department and "both [he] and Dewey were not averse to seeing established a band of younger teachers who would understand their lifework, be not unsympathetic to it, and perhaps even extend it in their own ways."[9] It seems no one was made assistant professor in the Columbia Philosophy Department who had not first graduated from the college and taught the course *Contemporary Civilization* made famous by James Harvey Robinson, and later, taught by John H. Randall, Jr. Richard Peter McKeon met these criteria and was made assistant professor in 1929.

At that time, Robert Maynard Hutchins was about the same age as Mortimer J. Adler. Hutchins was dean of the Yale Law School, while Adler was teaching psychology at Columbia. Adler and Hutchins became friends during a collaborative study on the rules of the laws of evidence. According to Adler, it was his idea that Hutchins should try to create a philosophy department at the University of Chicago to rival Harvard's great era of James, Royce, Santayana and Palmer. In an act of immodesty, putting the best possible face on it, Adler proposed that the process should begin with Hutchins' hiring him and his friends, Scott Buchanan, Richard McKeon, and V.J. McGill. It would later turn out that what Hutchins and Adler had in mind was nothing new but rather, a resurrection of the great ideas of the past, heavily infused with Christian theology. The blend was to be installed at Chicago as a firm curriculum.

It was the pressing perception that Chicago was slipping from its "earlier role as leader in American education" that had fueled Hutchins' nomination. The university needed a leader as "brilliant, brash, and bold" as Harper once had been.[10] If Hutchins qualified on those criteria he had a down side. There had been an acknowledged risk in hiring Hutchins into the top spot at Chicago, according to William McNeill, who had spoken with Robert A. Millikan, Distinguished Service Professor Emeritus at the University. Hutchins' "youth and panache were hard for his elders to accept." As McNeill characterized him, Hutchins was "stubbornly sophomoric, seeking truth, scorning compromise, and, every so often, impugning established authority with his characteristic barbed wit." Apparently, Hutchins was also a man of "extraordinary physical beauty." McNeill says,

> Without it he would not have risen as rapidly as he did....Men as well as women were attracted to him; and a sexual undertone...could not be excluded from many, perhaps most, [of Hutchins'] encounters with others....He often seemed arrogant, and the result was to keep ordinary, casual acquaintances very much at arm's length....To hold one's own in his intimate company, it was almost necessary to ...[resort] to raillery which sometimes became competitive as each egged the other on toward new heights of impudence.[11]

According to Gary Cook, it was "surprising" that Hutchins took Adler's suggestion to restructure the Chicago School seriously. After all, who was Adler in 1929? But Hutchins became enraptured. Hutchins seemed to operate from behind a beautiful mask, which separated his inner reality from the charmed, infallible fellow who brought chaos to the Chicago Philosophers. Before Hutchins and Adler, the success of Chicago Philosophy had resided in quiet, amiable, unassuming personalities, committed to democratic principles, openness, and philanthropy.

Robert Hutchins' half-hidden agenda, high-handed decision-making and back-tracking on promises was something these gentle men did not know how to handle.

The Chicago faculty in 1929 had almost all come from Protestant backgrounds. While they abandoned the dogmatic teachings of their youth, they remained strongly connected to religion in principle. McNeill says, "A strong evangelical residue remained. As most professors saw things, teaching was an improvement on preaching...." Teaching reason as the antidote to outworn beliefs became the Chicago secular ministry. Chicago was not prepared for Hutchins' high-handed administrative style, his self-important aura, his youth, or his passionate conviction to some new dogma of his own making (or Adler's making).

Just a week after Adler had made his brazen suggestion Hutchins wrote to Adler's friends to obtain their resumes. In October of 1929, although Hutchins was still not fully installed, he seems to have informed Tufts of his intentions to add people to the Philosophy Department. (Now it was Adler, Buchanan, and McKeon; McGill had dropped out.) At first, the Department tried negotiation. In the beginning, there was some support for the appointments. Gary Cook has suggested that it came from Smith, but there is evidence that early support came from Burtt as well, since he wrote the initial invitation to Adler to come to Chicago. Burtt's cooperation was soon to deteriorate in the wake of Hutchins' duplicity. Yet, T.V. Smith continued to enjoy a lively relationship with Hutchins both before and after the crisis, perhaps because he was away on leave at Cornell in 1930-31. Perhaps there were other reasons. Hutchins had written to Adler in October that the "most important member of the department, who is not the chairman, has enthusiastically endorsed my proposals." This person must have been T.V. Smith.[12]

Tufts discussed the possibility of Hutchins' proposal with the members of his department and summarized their responses in a three-page memorandum to President Hutchins, dated October 28, 1929. Cook's assessment is that Tufts was "diplomatic to a fault" in presenting the department's view.

Tufts had written:

> While we believe that the three men in question are men of scholarly promise, we are not prepared to say, without further consideration, whether

> if we were to take into account all the possibilities in Philosophy, we should settle upon the three men in question as the best available for the needs and purposes of the department.[13]

The Chicago Philosophers were not averse to hiring Richard McKeon, but were cooler about Buchanan and Adler.

Hutchins did not respond to Tufts' memorandum. It was late November. On December 4, still without responding to Tufts, Hutchins made his tactics known in a letter to Adler. "Our present plan is to retire two men and appoint you in their place...The financial situation is such that we probably can't swing Buchanan and McKeon this year. I have great hopes however that we can bring them in the year following."[14]

Hutchins was not wasting any time putting his plan into action. He contacted Tufts for the second time to ask if the department would accept Adler under the conditions that the appointment would not necessarily be permanent, but at the discretion of the department. Adler's "strong preference was to be closely connected with philosophy," but his teaching assignment was to be one-third in philosophy, one-third in psychology and one-third in law. [15] In a move they were later to regret, the department gave its approval. It was Burtt, whom Adler had known at Columbia, who wrote him that the "Philosophy Department ha[s] accepted [your] proposal to teach."[16] Adler described his appointment in his biography and cast Burtt as a "newer member" of the Chicago School, although he had been there for seven years.

Adler was brought in at a rank of associate professor of philosophy for three years with a salary of $6,000 per year. This was much more than the other Chicago philosophers were earning, and it was a huge jump for Adler. Adler had been earning only $2,400 at Columbia. His Chicago salary put him a full $1000 a year above anyone on the Philosophy faculty except for Tufts himself. Tufts urged that Hutchins should make adjustments in the other salaries to compensate, but Hutchins refused.

Adding insult to injury, rumors were circulating as far away a California that Adler had been speaking out against the Chicago department and especially its brand of philosophy. In and around Columbia, where he was still in the Psychology department, Adler was letting the cat out of the bag about his plan with Hutchins. By this time, Tufts began to doubt that Hutchins had confidence in him to lead the department since the rumors coming up from Columbia insinuated that Adler and Hutchins intended to change the personnel in the Chicago group radically.

Tufts resigned as chairman and considered retirement. This was a shock which some believed softened Hutchins to extend an olive branch. During the brief period of reconciliation Tufts agreed to teach on a limited basis in the 1930-31 academic year. George Mead, then sixty-seven years old, took over the chairman's position. He could hardly have imagined what was about to happen. Mead's year as head of the Chicago group was literally the death of him.

In the Fall of 1930, Mortimer Adler came to Chicago, but he did not teach any philosophy courses. Instead, he and Hutchins began the "Great Books" program with twenty selected freshmen students. They designated the new curriculum "honors program." Adler signed up for Hutchins' committee to consider reforms to the undergraduate study program. He spent little or no time in and around the Philosophy Department and the philosophy professors did not pursue him. Specifically, they did not ask him to their meetings and conferences, held at lunch time at the Quadrangle Club. The group was to be roundly criticized for this slight of protocol later. When Adler did go to department meetings, there was friction over what the philosophers perceived as the "elitism" of Hutchins' and Adler's so-called "honors program." This was the academic year 1930-1931.

In the Fall of 1931, since Tufts had retired, or nearly so, and Mead was making plans to follow him, the Philosophy Department felt the need to cast about for possible new faculty. They proposed bringing in Albert P. Brogan from the University of Texas. Hutchins had different ideas and proposed bringing in Scott Buchanan from the University of Virginia. Buchanan was to become an official member of the Philosophy Department but he would be teaching in mathematics and biology as well as in philosophy. Naturally enough, the Philosophy Department did not favor this proposal--for Buchanan was to be paid from the Philosophy budget. Like Adler's appointment, Buchanan's was supposed to be "not necessarily permanent." To allay the monetary question Hutchins declared that Buchanan's salary would be paid for three years from a source outside of the university.

Hutchins' proposal raised new fears that the agenda rumored from Columbia the year before was being pursued. Not surprisingly, the department asked Hutchins for clarification of his position on the Department of Philosophy and his future plans for them.

Hutchins replied. "His answer was an intimation that he did not know whether or not he wished to commit himself to a department in the future built up on the basis of its present membership."[17] Instead he informed the department of his intentions to poll noted philosophers as to the "relative merits" of the present members of the Chicago department, his own favored new picks and any individuals the department was intending to draft. This was an incredible thing to conceive in the first place, but worse, Hutchins carried it out. He prepared a questionnaire asking external evaluators to rank in order of merit a number of younger American philosophers, including Buchanan, McKeon, and some of the members of the Chicago department.

Mead had apparently submitted a list of names of individuals, whom the Chicago Philosophers deemed capable of performing such an evaluation. Hutchins did not show Mead his questionnaire, nor did he consult with him, but he sent out the invitation to evaluate to eight prominent people, four from the department's list and four of his own choosing. Hutchins did not see fit to show the replies to the members of the department. Ironically, it turned out that the replies were strongly supportive of the existing faculty. Hutchins' flagrant act angered even his own picks, Buchanan and McKeon, who resented being put on a list and evaluated behind their backs and without their consent.

Outrageous, high-handed and humiliating, the evaluation set off the tinder-box explosion that had been brewing since Hutchins came to the University two years earlier. Even as the evaluation procedure was underway, Mead had recommended that the Chicago Philosophers begin to look elsewhere. The resignations came in December 1931. On December 3, 1931 a protest, written by E.A. Burtt on behalf of professors, Mead, Murphy and Burtt, arrived at the secretary's office of the Faculty of the Division of the Humanities.[18]

Burtt had an offer from Cornell, Murphy had one from Brown University and Mead was invited to Columbia at an annual salary of $10,000, twice what he earned at Chicago. In the middle of January 1932, Burtt, Murphy and Mead issued an ultimatum. Mead brought it by word of mouth to the dean of the humanities division. They wanted Adler and Buchanan out of the department; they wanted salaries comparable to those paid to those two men; they wanted to offer an appointment at full professorship to Albert P. Brogan from the University of Texas. Last, they wanted a response from Hutchins by 10:00 am. the next day.

Hutchins met with Mead at 10:00 AM as requested and what transpired was summarized by Dean G.J. Laing, dean of the humanities division in a document titled, *The Situation in the Department of Philosophy*. Right after the meeting, Hutchins left for Arizona, where he remained with his wife until mid-February of 1932. During his absence Dean Frederic C. Woodward kept Hutchins informed of events as they developed. On January 22, 1932, Woodward sent a cable giving the impression that the situation was easing. Woodward believed that Mead and Murphy were intending to stay at Chicago, but Burtt was likely to leave. In less than a week (January 28), he cabled again with the news that Murphy was intending to take an appointment at Brown because he was distressed about the possibility of filling vacancies at Chicago considering the uproar. Murphy sought "a more congenial atmosphere" than the one Hutchins had created at Chicago. Woodward assessed Burtt at that time as "irreconcilable." Word that the whole thing had blown up came on February 6 when Woodward sent a final telegram to Hutchins.

> Morning papers carry news of resignation of Mead, Burtt, Murphy. We had fair success in preventing sensational story but faculties seriously disturbed.[19]

A summarizing account was written by Dean G.J. Laing, who strived to give the administrator's point of view. Laing was loyal to Hutchins, and his document is stereotypically middle managerial, but self-eviscerating in contrived excuses. The piece is interesting for its suggestions, between the lines, of the behind-the-scenes discussion that had gone on. The last straw was on the camel's back and the men of the Philosophy Department at Chicago University all left to go in separate directions.

Laing saw the need to go on a little about the value of "common sense" and his own support for the new direction Hutchins had taken at the University.[20] Except for the notoriety, Hutchins must have counted himself a lucky man to have so expeditiously engineered the clean-sweep.

Burtt's subsequent work suffered the consequences of the political in-fighting and petty politics at Chicago. It was some time before he recaptured the threads of ideas he had been working on during the Chicago years. It was the end of an era for him as well as for Pragmatism at the Chicago school. Burtt moved on to Cornell in 1931 where joined a group of older Idealist philosophers.

Mead died unexpectedly, following an illness on April 26, 1932. Murphy stayed at Brown until 1939, when he left to become an administrator at the University of Illinois. There, according to Professor Marcus Singer, Murphy's friend and a Murphy scholar, he "began a career as administrator and department builder."[21] In 1945, Murphy went to Cornell as department head, where he crossed paths with Burtt again. At Cornell, Murphy proceeded to restructure the Philosophy department along analytical lines, alienating Burtt and the other older philosophers who had given the school its Idealist reputation.

Burtt did not bring Chicago Pragmatism with him to Cornell. He joined a group of well-established, late-career men. In 1932, the Cornell Philosophy department was the country's center for Idealism and remained so until Murphy arrived in 1945. When Murphy left Cornell in 1953, it was the country's strongest analytical school.[22] Murphy brought in Max Black, whom he had known from the University of Illinois and Norman Malcolm, who came from Princeton. Also in the mix sometime after 1947 was Gregory Vlastos, from Queens in Canada, replacing George Sabine, who retired. Vlastos was a Greek scholar who grew to have a deep, personal friendship with Burtt. Vlastos once told John Copeland that "he [Vlastos] had come to have a lot of respect for Prof. Burtt, but that Prof. Murphy had not treated Burtt very well."[23]

Harold Smart, George Sabine, G. Watts Cunningham and Burtt were all in the department when Murphy arrived. After he arrived these older men were "pretty much in the discard," as graduate student, John O. Nelson, put it.[24] Burtt was the last to retire, and according to Bernard Wand, another graduate student, "it [serves] as an indication of Burtt's status in the department at the time that his office was one of the graduate seminar rooms."[25] These facts, help to explain why Burtt did not succeed in carrying the Chicago School of Philosophy with him to Cornell, although his own pragmatist philosophical interests, even as they developed and grew along religious lines, harmonized well enough with the Chicago tradition and the tradition of idealism at Cornell.

What had marked the Chicago school and distinguished it from the other great institutions of the era, Harvard, Yale, Columbia and Cornell, was the union of philosophy with the departments of psychology and education to form a Department of Pedagogy, complete with an experimental school and through Ames, a liberal church. According to Darnell Rucker in *The Chicago Pragmatists* (1969), the world view that William James celebrated as a school of philosophy in its own right, was concerned first and foremost with activity as a central concept. It had three important features.

> First, [activity] is a process theory; activity is something going on, and agents or persons and objects or the world are alike results of the process. Second, agents are essentially social beings; consciousness is the product of processes in which a number of agents interact, and third, ends are relative to the conditions of activity at a given time; a genuine evolutionary process is one in which real novelty enters, invalidating, preconceived, fixed goals of action.[26]

The major American Pragmatists are, of course, William James (died 1910), Charles S. Peirce (died 1914) and John Dewey, who out-lived the movement. Among them all, it was Dewey who did the most to put pragmatic ideas into the mainstream of educated citizens. Part of the Pragmatists' program was the belief that virtually anyone who was educated and living in a democratic society could advance human purpose in the world through open and honest inquiry. Pragmatism was establishing a new spiritual landscape in educated America, based on inquiry, democracy, utility, outspokenness, and common sense. Open, undogmatic, tolerant, and self-corrective, were the guide-posts marking a philosophy so spare that a whole range of diverse beliefs could be fitted into it.

Burtt's last important book, *In Search of Philosophic Understanding* (1965), is very much in this tradition. In that work Burtt raised open-mindedness to a philosophical method and demonstrated how such a method accommodated all the major philosophical idea systems of the entire world. In the end, that, too, was part of the problem, the Pragmatists over-generalized.

In important ways observed by T. V. Smith, Chicago Pragmatism was a house easily divided. Sometimes it left outsiders scratching their heads. Smith admired George Herbert Mead most, among all the Chicago philosophers. Yet, he wrote of him:

> Long after I had become a colleague of Mead, I asked him one day at lunch...whether he thought that there was anything existing before life came upon the scene. This seemed to me a question to be answered plainly Yes or No, depending upon one's drift toward Realism or Idealism. Mead answered the question at length...I repeated the question for a yes or no answer. He answered it at greater length....I never did understand; and...I came to doubt whether he did.
>
> I took it that he was confused, having left Idealism (Hegelianism) and not having arrived firmly at anything else. ...Confusion, between the knower, or the knowing, and the known, seemed so to dog the steps of the Pragmatists that I decided that they were all what I came to call "basement-Idealists" rather than, with Hegel, the "attic" kind. They all seemed to me to doubt--what I could not doubt--that anything existed apart from experience, and yet they seemed unwilling to face the consequences of such a position. They wanted to be Idealists without giving up the fruits of Realism. It made them unhappy to be thus accused, but so they seemed to me.[27]

I dislike leaving the last word on the Chicago School and its distinctly American philosophy with T.V. Smith, but the difficulties he describes with such felicity seem to have been genuine stumbling blocks for anyone outside of a special initiation into Dewey's logic. Kind men, open-minded and diplomatic to a fault, accepting and sincere philosophers, the American Pragmatists suffered from indefiniteness, a trait the modern analytic philosophers, with their dedication to precision, could not abide.

Pragmatism ended up being too shallow to accomplish what it set out to do; the high value it placed on liberation of all ideas turned into a vacancy of unifying ideas. Even well educated people are not ready to relinquish a ready-made and self-sustaining world-view, a difficult but necessary first step to accepting Burtt's Pragmatist alternative, individual thinking. To the extent that Pragmatist fundamentals have found their way into contemporary thinking as postmodernism, they have been divisive, rather than unifying, as the original thinkers intended.

In an article for *The Journal of American History*, *The Problem of Pragmatism in American History*, David Hollinger explains the "case of the vanishing pragmatist" as part of a larger historiographic phenomenon, "the case of the vanishing American." While the intention and hope of Pragmatism was to unite the culture behind the scientific habit of mind, democratic values, and a pluralistic theory of truth, the modern-day leitmotif of American historiography is not necessarily holism of method, but strong belief in the pluralism of experience; the search for one truth is abandoned. Hollinger observed that the classic studies of "Puritanism and of the political culture of democracy have reconstituted themselves in keeping with the insight...that Americans of different classes, ethnic backgrounds, or educations create and experience very differently the themes that still serve, somewhat lamely, to unify 'American History'."[28]

The Pragmatists' insistence on the reality of each individual's experience eroded the common ground of that philosophy. The one essentially "American philosophy" ultimately could not defend itself against all comers. It was too open-minded, too democratic. Pragmatism could not unify the culture. Making the truth plural, it diminished the value of truth and diminished itself in the process. Burtt's later program, developed in the late 1940s, to build consensus through synthesis of the world's major religions and philosophies through "cooperative evaluation" might be seen as one answer to this problem.

With the collapse of the so-called Chicago School, the influence of official Pragmatism diminished. Still, it has not lost all its vitality. Remarkably, the central ideas have filtered into many perspectives and operate surreptitiously in the scholarship of a variety of disciplines. Pried away from the original tradition, Pragmatism's themes are often the undercurrent for spirited controversy and debate. For one thing, the mood of the young radicals and their early pragmatism can be found in present-day postmodernism. For another, the liberal Protestantism, favored by the radicals, attracted a group of young men from Chicago who called themselves Religious Humanists in the 1930s. They contacted many of the radicals, including Burtt, to sign their "Manifesto" in 1933. Religious Humanism evolved into the movement we know today as secular humanism.

CHAPTER 6

RELIGIOUS HUMANISM

The movement calling itself the Humanist Fellowship emerged from among a handful of young (and some said rash or hasty) pragmatist thinkers, including a number of Unitarian ministers associated with the University of Chicago, in the early 1930s. They were mostly students from area universities and colleges. Members of the original group, living today, rightly claim to have been the forerunners of contemporary humanist organizations, such as The American Humanist Association with hundreds of chapters in the United States and abroad. Secular humanism, with its atheism and anti-religion stance is the descendent of the Humanist Fellowship, which, ironically, started out as a religious organization.

The small Chicago group became known first through its mimeographed newsletter, *The New Humanist,*[1] a paper, which grew eventually into the established journal, *The Humanist.* The lively, enthusiastic young men who wrote for the earliest newsletter undertook a publicity campaign for their organization, which they carried out through the mail, contacting many eminent writers and philosophers. Inspired by Dr. A. Eustace Haydon, professor of comparative religions at the University of Chicago, the first edition appeared in April 1928 and carried the heading *A Bulletin of the Humanist Fellowship.* In the beginning, Dr. Haydon paid for the mimeographing. Eventually Hayden became sponsor and editor of *The New Humanist,* which evolved from the bulletin.

Raymond Bragg, associate editor of *The New Humanist*, was a young Chicago-based Unitarian and student of Haydon. Edwin H. Wilson was a practicing Unitarian minister in Dayton, Ohio. Haydon, Bragg and Wilson were the chief organizers of the "Humanist Manifesto" project in 1933.

Writing a manifesto became a feasible project when Raymond Bragg enlisted Roy Wood Sellars, a well-known pragmatic thinker and author, and asked him to prepare a document outlining the basic tenets of belief for a first draft and then established a small editorial committee to refine it. The idea behind the project was to issue "a challenge in the name of naturalism to the supernaturalists whose beliefs were based upon revelation rather than reason and science."[2] Sellars was well-known, a professor of philosophy at the

University of Michigan at Ann Arbor, a religious liberal, and the author of the book, *The Next Step in Religion*. In that book, Sellars had used the term "humanist" in the last chapter.

The project organizers compiled a list of potential signers from among established Unitarian ministers and published authors and philosophers whom they imagined might be sympathetic to their program. They contacted Professor E.A. Burtt at Cornell University right away, sending him a draft of the manifesto written by Sellars. The same document was sent to John Dewey, John Herman Randall, Jr., at Columbia, Walter Lippmann of *The New Republic,* James Harvey Robinson, at the New School for Social Research in New York, Corliss Lamont, who eventually became associated with humanism as its foremost spokesman, and many others, about thirty in all. Dewey and Randall eventually signed the document, as did Burtt, although Randall's initial copy somehow got lost in the mail. Lippmann, Robinson and Lamont did not sign. All had been associated with Columbia University and the young radicals (except Lippmann) at one time or another. The framers were of such a mind at the time that they marked the first mailing "CONFIDENTIAL". [3]

Burtt must have been initially excited about the possibilities of advancing a philosophy of religion based on scientific method and scientific principles because he responded right away, making a profound impression on the group of young students. Burtt had argued for such a program himself in 1929 in his second book. This was *Religion in an Age of Science*, part of a series edited by John Herman Randall, Jr. Randall had intended that his *Religion and the Modern Age Series* be "a series of brief, clearly written books dealing constructively with the problems of present-day religious life in the light of modern knowledge and modern social conditions."[4] Randall and his father, John, Sr. had completed the first volume, *Religion and the Modern World*. Clearly, Burtt's contribution grew out of his friendship with the Randalls of New York City and especially John Randall, Jr., at Columbia, and Randall's respect for Burtt's early work, taking the pragmatists' view on modern science.

In *Religion in an Age of Science*, Burtt had argued against William James' theory of religious truth, known from *The Will to Believe*. This is James' notion that "when anyone faces a live, forced, and momentous option between two hypotheses that cannot be decided on intellectual grounds (by which James means, cannot be decided by appeal to overt physical data) he is justified in believing the hypothesis that is most satisfying to his emotions."[5] In 1929, Burtt rejected this notion as impossibly unscientific, taking belief in the existence or non-existence of God as an example. Harking back to the Humean design argument turned upside down in *The Metaphysical*

Foundations, he noted that the concept of God implies universality, just as scientific law requires it. If God is objectively real then all conscious beings must experience Him equally and everywhere as we experience gravity. On James' terms, belief in God is experiential and on an equal footing with non-belief. This admits experiences which cast doubt on God's reality and makes His non-reality a possible reality of experience in the universe. To recognize the experience of doubt then, is to negate the universality of God and thereby to negate what is meant by the concept of God in the first place. The argument concludes:

> God's universality [thus, reality and existence] can only be verified by his universal discovery; since that discovery is not actual in the case of all men, I cannot properly affirm the validity of the concept no matter how strongly I feel certain of it myself. I can only progressively reveal its validity by removing the obstacles that stand in the way of its appearing valid to others, and this is a matter of moral action, not of theological assertion. To do otherwise is to make oneself logically comparable to a scientist who should insist on the universal validity of a discovered law with which experiments of others were still in conflict.[6]

This is what attracted the Humanists to Burtt. When the *Humanist Manifesto* appeared in print, signed by thirty-four well-known intellectuals and religious thinkers, including Edwin Arthur Burtt, John Dewey, and John Herman Randall, Jr. Sellars wrote a companion article in which he explained:

> In the Humanist Manifesto it will be seen that many of us have reached a common body of beliefs and attitudes, beliefs about man, his place in the universe, the general nature of that universe, and attitudes toward the great questions of life...[7]

On April 10, 1933, Burtt wrote back to the editors of *The Humanist Manifesto*. The tone is precautionary.

> I have just received the copy you kindly sent me of the projected Humanist Manifesto. I beg of you with all the earnestness at my command not to publish this statement without further consideration of its implications. It is not that my own agreement and signature are of any consequence, but I am sure that a manifesto in this form will distress and alienate from the humanist movement a large number of people whom it is not at all necessary to alienate. It is quite possible that this public pronouncement, signed by the persons whose support you contemplate securing, will mark a historic landmark in the development of religious humanism, and will be accepted by the public for what it claims to be, namely a creedal definition of the essentials of religious humanism whose major implications ought to be accepted by anyone who proposes henceforth to call himself a humanist and wishes to cooperate fraternally in his religious life with other

> humanists. For this reason it seems to me that it would be a tragic calamity not to make sure with the utmost care and caution that the pronouncement opposes only those positions and ideas which are irreconcilable with the essential matters on which those in profound sympathy with the humanist movement take their stand. Just at the moment I am absorbed in other duties that cannot be put off, but in a day or two I shall venture to write again, making specific suggestions which in my judgment are necessary if the unfortunate consequences above mentioned are to be avoided. Briefly, it seems to me that a natural reading of this statement would assume that it commits itself to a particular theory of naturalism, excluding all other naturalism such as the Aristotelian, which would allow a certain metaphysical reality to teleological relations, irreducible to casual connections of the material and genetic types. It would assume that the humanism denies the reality and religious value of all entities transcending human experience, whereas, if I have read my humanist friends correctly, all that they mean to insist upon as essential is that if such entities are accepted their meaning and value for us may be constructed in terms derived solely from human experience. It would assume that humanism denies the legitimacy of carrying over terms (such as God) from the older religious framework, whereas all that is needful to insist upon is that if these terms are carried over they must be fully and honestly reinterpreted in terms consistent with scientific truth and shareable human values.
> Please do not make the irreparable mistake of letting this go quite yet.
>
> E.A. Burtt[8]

Raymond Bragg replied immediately and noted how anxious he was to receive the full extent of Burtt's recommendations. Burtt wrote again on April 12, before receiving Bragg's letter, dated April 13. This time Burtt sent the original document with changes specified in his handwriting. The document is reprinted here with Burtt's changes in brackets. It can be seen from the footnotes and bold-face-type notations within the draft which of the suggestions Burtt made were ultimately adopted or rejected.

> (Confidential Draft)
> A Humanist Manifesto
>
> The time has come for widespread recognition of the radical changes in religious beliefs in the modern world. The time is past for mere revision of traditional attitudes. Science and economic change have disrupted the old beliefs. Religions the world over are under the necessity of coming to terms with new conditions created by a vastly increased knowledge and experience. In every field of human activity, the vital movement is now in the direction of a candid and explicit humanism. In order that religious humanism may be better understood we, the undersigned, desire to make certain affirmations which we believe the facts of our contemporary life conclusively demonstrate.
>
> There is great danger of a final, and we believe fatal, identification of the word *religion* with doctrines and methods which have lost their

significance and which are powerless to solve the problem of human living in the Twentieth Century. Religions have always been means for realizing the highest values of life. Their end has been accomplished through the interpretation of powers in nature, and the creation of a technique of worship. By means of this worship men hoped to exercise control over [*win adjustment to*] those powers in order to attain values considered by the group to be most desirable.[9] A change in any of these factors results in the alteration of the outward forms of religion. This fact explains the changefulness of religions through out the centuries. But through all changes religion itself remains constant in its quest for abiding values, an inseparable feature of human life.

Today man's larger understanding of the universe, his scientific achievements, and his deeper appreciation of brotherhood have created a situation which requires a new statement of the means and purposes of religion. Any such vital, [*such a*] [10] fearless, and frank religion capable of furnishing adequate social goals and personal satisfactions may appear to many religious people as a complete break with the past. While this age does owe a vast debt to the traditional religions, it is nonetheless obvious that any religion that can hope to be a synthesizing and dynamic force for today must be created for the needs of this age. To establish such a religion is a major necessity of the present. It is a responsibility which rests upon this generation. We therefore affirm the following:

First: Religious humanists regard the universe as self-existing and not created [*formed*].[11]

Second: Humanism believes that man is a child [*part*][12] of nature who has emerged as the result of a continuous process.

Third: Mind is a function of the organism. The traditional dualism of spirit and body must be rejected. [*Mind and body are closely interconnected.*] [13]

Fourth: Man's religious culture and civilization, as clearly depicted by anthropology and history, are the product of a gradual development due to his interaction with his natural environment and with his social heritage. The individual born into a particular culture is largely molded by that culture.

Fifth: The nature of the universe depicted by modern science makes unacceptable any supernatural or cosmic guarantees of human values. Religion must formulate its hopes and plans in the light of scientific procedure.

Burtt Proposed:

[*In denying the supernatural humanism does not deny the existence of realities transcending human experience. But it insists that the only dependable way of determining the meaning and value for us of such realities, as of any others, is by honest study and intelligent assessment of human experiences of value realized in relation to such realities.*] [14]

Sixth: We assert that the time has passed for theism, deism, modernism, and the several varieties of "new thought." [*If terms such as God, salvation, soul, and the like, are to be retained in humanist thought, they must be reinterpreted without reservation in terms of verifiable scientific knowledge and empirically discoverable human values.*] [15]

Seventh: Religion consists of those actions, purposes, and experiences which are humanly significant. Nothing human is alien to the religious. It includes labor, art, science, philosophy, love, friendship, recreation--all that is in its degree expressive of intelligently satisfying human living. The distinction between the sacred and the secular can no longer be maintained.

Eighth: Religious humanism considers the complete realization of human personality to be the end of man's life and seeks its development and fulfillment in the here and now. This is the explanation of the humanist's social well-being.

Ninth: In place of the old attitudes involved in worship and prayer the humanist finds his religious emotions expressed in a heightened sense of personal life and in a cooperative effort to promote social well-being.

Tenth: It follows that there will be no uniquely religious emotions and attitudes of the kind hitherto associated with belief in the supernatural.

Eleventh: Man will learn to face the crises of life in terms of his knowledge of their naturalness and probability. Reasonable and manly attitudes will be fostered by education and supported by custom. We assume that humanism will take the path of social and mental hygiene and discourage sentimental and unreal hopes and wishful thinking.

Twelfth: Believing that religion must work increasingly for joy in living, religious humanists aim to foster the creative in man and to encourage achievements that add to the satisfactions of life.

Thirteenth: Religious humanism maintains that all associations and institutions exist for the fulfillment of human life. The intelligent evaluation, transformation, control, and direction of such associations and institutions with a view to the enhancement of human life is the purpose and program of humanism.

Burtt Proposed:

[*Humanism expects that religious associations will pass through a progressive remolding, as experience teaches the best methods by which humanists may cooperatively further their common aims.*][16]

Fourteenth: The humanists are firmly convinced that existing acquisitive and profit-motivated society has shown itself to be inadequate and that a radical change in methods, controls, and motives must be instituted. A socialized and cooperative economic order must be established to the end that the equitable distribution of the means of life be possible. Humanists demand a shared life in a shared world.

Burtt Proposed:

[*The ultimate goal of humanism is the universal cooperation of mankind in intelligent pursuit of the common good.*][17]

Fifteenth and last: We assert that humanism will: (a) affirm life rather than deny it; (b) seek to elicit the possibilities of life, not flee from it; and (c) endeavor to establish the conditions of a satisfactory life for the many, not merely for the few. By this positive morale and intention humanism will be guided, and from this perspective and alignment the techniques and efforts of humanism will flow.

So stand these theses of religious humanism. Though we consider the religious forms and ideas of our fathers no longer adequate, the quest for the good life is still the central task for mankind. Man is at last becoming aware that he alone is responsible for the realization of the world of his dreams, that he has within himself the power for its achievement. He has only to set intelligence and will to the task.

The most critical points for Burtt were the mind-body problem in thesis three and the semantic problem in thesis six. Neither of these were adopted, nor were they adequately addressed by the editors in the final draft. Randall had objected to Burtt's statement on the mind/body question in thesis three and Robert M. Lovett, a newspaper editor, had objected to the semantic issue in thesis six.

In his second letter to Bragg, Burtt had written again admonishing care "in preventing the alienation from the humanist movement of a group of people whom there is no need to repel." After indicating the minor changes he was recommending, he went on to say:

More important matters are, in my judgment, the following. The first sentence of the third thesis, as it stands, might seem to claim conclusiveness on a matter which science has not conclusively decided, and to commit humanism to the old-fashioned sort of materialism....

This was followed by the changes he proposed should be adopted between the fifth and sixth theses and his changes in the sixth thesis.[18] Randall, receiving his draft late, received Burtt's suggestions along with it. Randall fired back, "I object to Burtt's change in the third thesis, other changes acceptable. I very much dislike the crass optimism of the last two sentences of the Manifesto."[19]

As to Burtt's semantic recommendations, Robert Lovett took umbrage. He wrote to Bragg on April 20, 1933:

> I have looked over the Humanist Manifesto and made no changes, except scattering a few commas. Except in article 6. I think it rather absurd to instruct people as to how they shall use certain words. These words have a metaphorical value in literature, and I should say it was impossible to use them to symbolize "verifiable scientific knowledge," etc. I think this sentence unfortunately naive, and tending to color the whole document. It will certainly be seized upon for destructive comment. Personally, I cannot give up God in such expressions as "God damn it all!" any more than I can give up liquor.[20]

Bragg sent a copy of Lovett's comments to Burtt on April 21, adding, "It is really quite funny and I am aware that you know Lovett." Burtt replied the next day and addressed his primary concerns in the fifth and sixth theses.

> Lovett's comment on the sixth thesis is certainly an illuminating illustration of the impossibility of telling in advance what words are going to mean to anybody. What I had essentially in mind in the modifications I suggested in the fifth and sixth theses was that there are many liberal religionists who still feel that they can give genuine meaning to traditional terms that have been associated with theistic or at least modernistic points of view, but who are yet entirely at one with humanism in what I take to be its decisive emphases, such as unreserved commitment to scientific method, determination of value by intelligent assessment of human experiences of good, and an uncompromising stand on the social and economic problem. It seemed to me important that any such should be welcomed and not repelled by the manifesto. I do not agree entirely with such people, but at the same time I am not committed to an extreme naturalism, and think it would be a great pity for the manifesto to imply more radical doctrines in this direction than is necessary for an unreserved stand on the things that humanism does count vital. That you and your friends agree with this as a general policy I assume from your kindly response to my previous letter.
>
> How would it do to restate the sixth thesis as follows: "We believe that the time has passed for theism, deism, modernism, and the several varieties of new thought. At the same time we welcome religious fellowship with any who interpret the human quest for the highest values in terms which have been associated with these positions (such as God, salvation, soul, and the like), as long as the interpretation is consistent with insistence on scientific method as the only sound guide to truth and on intelligent, democratic criticism of human experiences of good as the only dependable guide to ideals of value...."[21]

Burtt's grand program to develop a scientific metaphysics of categories rings out from this letter. He intended to restore the value to "theistic" terms by comparing them to scientific categories and dressing them in scientific descriptions of how the mind catalogues experience. Bragg had used the word "funny" to describe Lovett's criticism, referring no doubt to the line about liquor, but what was peculiar was that Lovett seemed to miss the point of Burtt's attempt to include believers, while guiding humanism to allow words associated with religious concepts to be considered "scientific". He wanted the Humanists to leave the door open for religious experience to enter the scientific world-view.

Many years later, remembering the history of the drafting of the Humanist Manifesto, Wilson wrote, "On the basis of this important comment by Dr. Burtt, some final changes were made." As can be seen this is not a fair statement. Burtt's semantic concerns and attempt to soften the positivistic bent of the manifesto were rejected. Still, it was Burtt's complete criticism and early attention to the young humanists and their document which inspired them and helped to convince them that they were on to something important. Unfortunately, as a group, they did seem to be out of touch with important philosophical concerns of the day such as semantics and the mind-body problem.

In 1929, Burtt had addressed the mind-body problem in an essay for the series *Man and His World* in the volume titled, *Mind and Behavior*. The essay is noteworthy not only for the position Burtt argues on the question, but also because it is the only time Burtt ever addressed the issue thoroughly in print. Burtt makes a case for mind as an emergent feature or characteristic property of some forms of life and life as an emergent feature or characteristic property of some forms of matter. The fallacy of materialism, the belief that the real is matter alone, is that "it assumes a complete kind of knowledge about a certain kind of situation, whereas in fact we are only in the process of getting such knowledge, and at no time is it fully attained." Likewise, the fallacy with idealism is that it "assumed as ultimate a timeless purpose, whereas experience reveals every purpose as a temporal process." Reminiscent of the paper, *Real vs. Abstract Evolution*, delivered at the Sixth International Congress of Philosophy at Harvard University three years earlier, Burtt wrote, "As actually revealed...the world of our experience is a process, not a finished thing, whereas in the traditional cosmologies [materialism and idealism] there was a tacit assumption that it is complete."[22] He argued that reality is a process and that "emergent evolution" typifies the process. New levels of complexity emerge from lower levels. "The more complex fact...reveals immediate qualities and modes of relatedness to other facts that were not present in the simpler behavior of the elements before they come together to constitute the complexity in question...The process...is one of the creation of novelty, as well as of the maintenance of continuity with the past; and the novelty takes the definite form of the appearance of more complex entities, with qualities and relations transcending those predictable in terms of the simpler conditions out of which they have emerged."[23]

> This way is to postulate, even at the lowest stages, an effective push, or drive, or nisus toward the higher qualities, so that, for example, the age of matter must be thought of as not merely an age of sheer mechanism, but as mechanism somehow working toward life, consciousness, and divinity. In just what form these higher qualities exist to be worked toward, and what part they play in their own progressive embodiment, are questions more often side stepped than frankly asked.[24]

At last, Burtt concludes: "Mind and matter thus understood are not metaphysically separate substances, but factors that function in every event occurring at every level."[25] Through this conception of mind and matter there is a definite place for experience, consciousness, feeling, value and spirituality situated within the scientific realm, considered as emergent properties of the increasingly complex levels of matter. Burtt wanted to leave space for "purpose" in the emergent properties of the universe, just as he had explained in his revised final chapter in the 1932 edition of *The Metaphysical Foundations.*

In responding to and guiding the young humanist fellowship, who had sought him out, Burtt made serious attempts to gain consideration of pressing philosophical questions he believed to be essential to any developing empirical, naturalistic or scientific religion. He wanted to consider the mind-body question scientifically and he wanted to leave "God", "salvation", and "soul" on the table in order to do this. I do not think that he was understood. Rather his interest was seized by less-adequate thinkers, who pushed the document into print. They interpreted Burtt's hearty interest as a sign of the excellence of their ideas and as a signal that the movement had some promise. They wanted to make newspaper headlines more than they wanted to consider underlying philosophical dilemmas.

Wilson writes of Burtt, "Dr. Burtt's interest in Western scientific and religious humanism diminished some years later, following his studies in the Orient and his considerable interest in Buddhism. However, his contribution to the evolution of A *Humanist Manifesto* was substantial...."[26] Well, yes from the perspective of the initiators, including, of course, Wilson. But of philosophical substance, they took little of what Burtt offered them.

In the end, thirty-four people signed the *Humanist Manifesto I.* Five were professional philosophers, Burtt, Randall, Dewey, Sellars and Oliver L. Reiser, Associate Professor of Philosophy, University of Pittsburgh. Nine were ministers, seven of those Unitarian. Fifteen of the entire thirty-four were Unitarians.[27] Robert Lovett, Editor of *The New Republic*, signed, joining three other editors, or former editors, of newspapers. As mentioned earlier, Walter Lippmann did not sign and did not even respond to the invitation.

A major spokesperson for humanism, Corliss Lamont, writing in 1974, demonstrates just how little Burtt's ideas had penetrated into the humanist tradition. Lamont states that the general philosophy of humanism is "materialism, stressing the fact that the foundation stuff of the universe is matter in motion. The second "strand is anti theological and anti supernatural, but it incorporates...many of the ethical teachings of outstanding religious leaders." "Third and most significant is the scientific. The Humanist insists that his inclusive philosophy must be consistent throughout with the established facts and laws of science." Fourth is that death marks the end of the conscious human individual and there is no personal immortality. Fifth is the idea of democracy and democratic procedures as key to the welfare of all mankind.[28]

While the third and fifth strands were something Burtt could easily support, he would have deferred comment on the fourth as outside of experience and therefore outside of human understanding altogether. The first and second strands would have repelled him. For Burtt, the core of human experience is both material and immaterial. To reunite them in some scientific way was his highest aspiration for philosophy. For Burtt, Humanism meant a philosophical method through which humans can come to understand themselves in the context of their experiences with each other and the universe. During the early 1930s, when the Humanists began to organize, Burtt was working out his own science of metaphysics, a theory of perception and cognition based on a study of mental categories which he argued had their roots in biological evolution. He presented his theory as a logical proof in *Genesis of Hypothesis*, an unfinished project he had formally begun just after arriving at Cornell.

CHAPTER 7

BURTT'S NEW PHILOSOPHY OF MIND

CHALLENGING THE COMMON SENSE CATEGORIES: TIME, SPACE, AND QUANTITY

Religious humanism was one reflection of the heavy burden the early twentieth century brought to the spiritually inclined. Sensitive intellectuals felt a responsibility to guide the culture toward a workable synthesis during the transition from the old faith to the new science. Truth standards based on belief had to make peace with scientific standards for certain truth based on strict empirical criteria. But the reconciliation that has gone on has not been all one-sided. Science has learned a thing or two about itself in the process and has backed away from claims to scientific certainty in all but the most technologically based fields. Among those who recommended self-awareness and humility for the sciences while pursuing a private path to the spirit was E. A. Burtt. Along the way, he advanced many of the ideas and methods nowadays thought of as postmodern. For him, they had their roots in responsible skepticism, modeled on David Hume, rational ethics, modeled on Spinoza, and an objective examination of the processes of mind, modeled on Kant.

Ever since *The Metaphysical Foundations of Modern Physical Science* appeared in print in 1924 philosophers and historians of science have pondered the relationship between phenomena as science describes them and phenomena as they are in themselves. Burtt showed that the Newtonian world-view was but one episode in a long parade of shifting patterns of thinking about objective reality and our mental constructions. His account of early modern science was written as the history of an intellectual shift. The Newtonian synthesis (as above, so below) was the impetus for the cultural and religious shift in man's view of himself in relation to God and the world, as well as for the modern philosophy of mind which accompanied it. Burtt's account of Newton could have raised serious doubts about what had been taken for granted as scientific objectivity since Newton, or more correctly according to Burtt, since the Philosophes corrupted Newton's philosophy with their own views. Beneath his historical account of modern physical science, Burtt was asking questions of enormous postmodern significance. Can the historical study of the changing patterns in world-views bring us any closer to an understanding of objective reality, if there is one? Can we come to understand our human condition and our relation to the world, if it exists apart from us, any better by studying our changing scientific descriptions of objective reality?

Today, at the beginning of the twenty-first century, these questions are entertained within a different context than when Burtt first opened them in the 1920s. Now we have had the benefit of V.W. Quine's *Two Dogmas of Empiricism*, followed by Donald Davidson, who demonstrated the errors inherent in the analytical philosophies and, lately, Richard Rorty, who has assigned little else to meaningful philosophy except to "continue the conversation of culture." We have had the benefit of the wide distribution of the ideas of Michel Foucault and J. Derrida. We are in the post-scientific age where deconstructing science is, at least, a legitimate enterprise, if not a commonplace one.

An important turning point took place in 1951 when Quine's work, *Two Dogmas of Empiricism*, came out. Quine undermined analytic philosophy by questioning the distinction between analytic and synthetic truth in an argument which demonstrated the circularity of the analytic method with sufficient rigor and clarity that the analytics themselves had to sit up and take notice. Quine can be placed in the same tradition as Pierre Duhem, John Dewey, E. A. Burtt and R.G. Collingwood in overturning our hopes for certain truth. He has been followed by Donald Davidson and Richard Rorty, who have stressed the Pragmatic elements of context-dependent interpretation of experienced phenomena. Although the so-called neo-pragmatists, acknowledged with full places in the contemporary debate, have down played metaphysics as such, they have something in common with Burtt. For example they rely on the spirit of the natural sciences to guide them; they relish exposing the presuppositions of empiricism and have recovered the big questions for philosophy by talking seriously about reality.

The Baconian dream, power over the environment, comfort, and material goods from technology, has been realized. Caught in a tidal wave of riches unimaginable to Bacon, or even to Burtt, not many thinkers have been circumspect enough to speculate about the truth value of the kind of knowledge science provides. It is pragmatic truth, after all, not absolute and not spiritually significant. It works to produce new inventions and new wealth whether or not it has any correspondence with the world as it is in itself, outside of human understanding of that world.

Burtt was different from most other philosophers of his day; he stayed the course of classical philosophy. He wanted philosophy to continue to ask questions about the meaning of human action and the purpose of man in the world. For him, describing the relationship between human experience of the world and the world as it is in itself was key to exposing the limits of science and preparing scientific philosophy to reach for an even higher level of understanding. This would be a new window on the spirit.

Over the years Burtt's questions have been steadily pushing forward in various fields, the history of science, philosophy and religion. What is mind in itself, if it exists in itself, and what difference, if any, does mind make to what is known? Should we give up the quest for scientific certainty, at least temporarily, in order to better entertain that question? Should we, and can we, give up theories of correspondence between science concepts and physical reality in order to really and truly open that question? If we study our own mental constructions, will we be in a better position to understand the world apart from us as it really is in itself? Can our awareness of how we experience and interpret physical reality open doors for spiritual understanding as well?

Clearly phenomenal worlds are not static. From culture to culture and age to age explanations of natural phenomena have changed. If all the phenomenal worlds could be studied from the historical perspective simultaneously, could relative truth value be assigned to any one of them? How shall we determine which aspects of those worlds are independent of mind and which are mental constructions only? One way to do this might be to subtract the anthropological basis for constructing the world, determine which capacities for constructing it are innate and which are learned by interaction with the environment.

Burtt took the position that the simple need for survival in a hostile environment has favored the natural selection of certain mental capacities at the expense of others with less immediate survival value. It is these favored capacities which structure and order our experiences leading us to believe that a certain kind of objective reality exists, universal order for example, when in fact, it is human perspective alone which we can know and which we mistake for objective physical reality. Burtt thought that when human beings have mastered the environment human evolution will take the next step. That will be intellectual evolution, guided by philosophers.

In an unpublished manuscript, *The Genesis of Hypothesis*, probably written in 1935, Burtt explores these possibilities. He took the first step toward answering the question, what is perception and how does it relate to objective reality? How do scientific hypotheses form out of the vast array of possible explanations? Why does a whole culture adopt certain explanations and dismiss others? What are the most significant operating categories of mind and how did we come to use those categories to invent a particular world-view?

It is worthwhile to trace these ideas as they developed in Burtt's remarkable mind. Sensitive to conflicts both within himself and within the culture, Burtt's philosophy of science and his philosophy of mind reflect the tension between modern science and the human spirit. Keeping the possibility of some positive reconstruction of the spirit in mind, Burtt wanted to reinterpret the methodological foundations of science to see if such a critique could lead to a new vision of both science and God.

After *The Metaphysical Foundations*, Burtt extended the ideas he had presented there. He challenged the common-sense definitions of science terms which have come down to us from the Newtonian world-view and through which we moderns make sense of our experience as we interact with the physical world. His approach was radical and completely fresh. In two papers, one read before the Sixth International Congress of Philosophy at Harvard University in 1926, *Real versus Abstract Evolution*, and the other, *The Contemporary Significance of Newton's Metaphysics*, published in a memorial volume on Newton in 1927, the framework was built for Burtt's grand scheme, a new philosophy of mind. In outline, he expected to reconstruct metaphysics into a scientific study of mental categories or universals, which shift from culture to culture and age to age. In a variety of arguments published over the next ten years he deconstructed space, time, causality, motion, quantity and logic, as simply categories, deriving from the scientific world view, and noted their origins in the history of thought by tracing transformations of their meanings through time.

Two years after *The Metaphysical Foundations* was published in England, Burtt reviewed A.J. Snow's *Matter and Gravity in Newton's Physical Philosophy* (Oxford Press, 1926). He opened his discussion with an opinion remarkable for its anticipation of post-modernity. Burtt saw himself as standing on the threshold of a new era in the philosophy of science.

> Signs multiply that physical science has passed from a dogmatic and deductive to a critical and reconstructive era. It is curious to note that when the words of a great genius [Newton] are taken to be absolute truth, no one proposes to investigate the genesis of his ideas, but once let his authority be systematically questioned, and it will not be long (especially in an age dominated by the notion of evolution) before historical analysis of his mental development will appear.[1]

Although he misread the positivistic direction philosophy and academic history of science were about to take over the next thirty years, he foreshadowed present-day trends.

Einstein's theory of relativity was still new when Burtt reviewed Snow's book on Newton, but many people could see its implications for philosophy. In September 1926, Burtt had his first opportunity to address an audience of professional philosophers. His paper, Real *versus Abstract Evolution*, was an attack on the realists' concept of Newtonian absolute time.

> We are told that all arises out of a matrix or primary level which is called Space-Time, or a whirl of electrons, and while we are wondering how these things of their own power can produce anything resembling the observable world, we discover that our metaphysician has kept a hidden solvent of all puzzles up his sleeve. We learn that he means by Space-Time or electrons not what the [Newtonian] physicist means by these things, but magic entities equipped from the start, like the scholastic occult causes, with the highly complex ability to do precisely what we need to have them do.[2]

This was the only indication that Burtt's talk would have a "post modern" theme. The realist analysis that followed was all showman-ship, delivered with the kind of conviction that he had adopted for speaking from the pulpit. Naturally gifted in the art of elocution Burtt left no wonder why he had won prizes for rhetoric and debate at Yale or why *The World*, a New York newspaper, had published a sermon he delivered at St. Paul's Methodist Episcopal Church on the Fourth of July, 1920 in its entirety.

For most of the hour he insisted on the reality of time as the literal experience of time. The order of discovery of facts in chronological time, as directly and empirically given alone, can be real. Toward the end he did an about face. Empirical time as real time is a ridiculous concept. All the while he was making the realists' case he was setting the stage for the idealists' point of view. Common sense shows that conscious experience is an integration of separate factors taken out of sequence and fashioned by mind into some seamless unity. Time only makes sense when events have been processed by thinking and ordered into a meaningful pattern. The point is that there is more to reality than literal, piecemeal reductionism to the logical simples of modern science just as there is more to the meaningful concept of time than isolated, disconnected events. The role of philosophy in intellectual life is to explain experience and the sense that minds make out of it as an integrated whole, not to concentrate exclusively on either the simple parts or the integrated whole, the way the reigning philosophies of the realism and idealism do. The goal of a philosophic understanding of science is to illuminate mental function, how the parts become the whole, which is to inquire into the processes of mind.

When experience is taken as a key to the interpretation of everything, common perceptions as universal categories and the fundamental logic of what presupposes what, taken together as metaphysics, is generated. It was the metaphysics of universal categories as a study of how mind uses universals as instruments that Burtt wanted to bring into the arena of philosophic discussion. It is quite possible that Burtt was inspired by the work of Ernst Cassirer. He had listed Cassirer's *Das Erkenntnisproblem in der Philosophie und Wissenschaft der neueren Zeit,* and two others, in the bibliography of *The Metaphysical Foundations.* Cassirer's work was not translated into English as *The Problem of Knowledge* until 1950. Like

Cassirer, Burtt believed that the guiding principles through which mind structures experience are not static, but evolving. Like Cassirer, Burtt was interested in making the leap from Kant's analysis of the organizing principles of natural science and morality to more all-embracing principles of the human mind in all its aspects. Like Cassirer, Burtt was interested in, or became interested in, developing a critique of cultures based on the organizing principles or categories he hoped to identify.

Then, too, Burtt might have taken some additional inspiration from the work of Pierre Duhem. Duhem's *L'evolution des theories physiques* (1896) is also listed in the general bibliography to *The Metaphysical Foundations* also. As both a philosopher and historian of science, Duhem stressed continuity over change. In a later work, *The Aim and Structure of Physical Theory* (1906) he separated metaphysics from science, at the same time attempting to save metaphysics from scientific objections. Duhem was sure that metaphysics is part of the explanation of physical reality, but that "no metaphysical system suffices in constructing a physical theory." While physics was subordinate to metaphysics in Duhem's mind, "metaphysics was beyond complete human knowledge...At the root of the explanation [physics] claims to give there always lies the unexplained."[3] For Duhem, as for Burtt, the reality of the physical universe in itself is beyond our grasp.

The thinking of Cassirer and Duhem bore new fruit for Burtt when he adopted the realists' own program of reducing complexities to logical simples and ferreting out metaphysical presuppositions. During this radical period Burtt was ruthless in his examination of all established ways of thinking. For some reason, he did not feel free to simply put his plan forward. He entered the discussion by deconstructing and disrobing the reigning philosophies. Vigorous, daring, and at his best, he coupled sarcastic good-humor with courage and wit. "I maintain earnestly that *the order of discovery is the order of reality, if we mean by reality something that can be empirically denoted and verified,*" he insisted.[4] The mood of his talk changed so subtly that the audience must have been confused when Burtt set them straight abruptly. "Beware lest these wild fancies bring philosophy into further disrepute with sober men of science! The time is surely come for you to pronounce this all a joke, and join in the laugh at its paradoxes...If we must go realist, let us go realist all the way."

> The point here maintained is simply that *no past event becomes an empirical reality until it has taken its place in the order of discovery. For that in the end is what empirical reality must mean.*[5]

We know by commonsense experience that mind is capable of constructing a time-line reality different from the one literally experienced.

For example, a person learns of his birth much later than he actually experienced the event. Still, he can construct a personal reality out of facts placed in time according to the conventional framework. Think of the geologist who learns of the vast transformations of the pre-human ages after he experiences the events of his own childhood. "Will he lend a hospitable ear to the notion that the vast transformations of the pre-human ages really happened after those of his childhood?"

The literal experience of absolute time (real time) is less important to human beings making meaning out of their lives in time than the construct of time (abstract time), which grows out of the ordering of events in a useful way. Burtt summarized, "Abstract evolution is the objective order that emerges when the living problems of the various perspectives have been solved and only the solution retained. Real evolution includes it and more--the story of how that objective order emerged out of the vagueness, uncertainty, and doubt which preceded."[6]

Burtt's underlying subject was his new conception of metaphysics, not Newtonian absolute time versus Einstein's concept of the space-time continuum, with which he had opened his remarks.[7] He wanted to understand how we make meaning out of experience using time as an instrument.

> Our question is of real time [literal experience] and its relation to abstract time [meaning constructed from time as a mental category]. On that question no science has any contribution to make, except the science of genetic psychology. There we will find profound illumination if we care to look. The point here maintained is simply that *no past event becomes an empirical reality until it has taken its place in the order of discovery. For that in the end is what empirical reality must mean.* [8]

Pushed to the limit of its logical consequences, empiricism, at least as it applies to the category of time, does not hold up.

The philosophical technique of pushing an idea to the limit was a favorite one as he later admitted in a paper for *The Philosophical Review* in 1953.

> One [of the most profitable procedures] is to take some promising philosophical idea and see what happens when it is pushed to the limit--i.e., when it is applied to all sorts of problems many of which had not been specifically envisioned when it was first adopted....Thus are the creative ideas elaborated which lead to new, clarified, and fertile interpretations in the sciences, in the arts, in statesmanship, in religion. The other way is revealed when thinkers become concerned about the conflicts that arise between what seems a promising idea to one philosophical school and what seems a promising idea to another school, and seek some way of analyzing such conflicts.[9]

Mind brings something more to the experience of time than strict empiricism can account for. There is association and recall for example, learning and ordering of events out of the literally experienced sequence. "The only world which we can be interested in and think about is the world that takes form and reality in the course of our experience"--that is, the one we construct in our minds. The conclusion is clear. Empiricism must confront actual experience, recognize the processes of mind which modify sensory input as it becomes conscious, and make the necessary accommodations in a new philosophy of mind. On this, Burtt had his own ideas.

Time is a metaphysical category, lying between the objective experience of physical reality and a constructed consciousness of meaning. Without metaphysical categories our experiences form no patterns and establish no context for our future experience. Burtt's innovation was that metaphysics, abandoned by the new realism, should become the new philosophy of mind. It would overcome the single-minded conception of reality adopted after Newton, that the primary qualities alone characterize nature simply because they are the ones most adaptable to mathematical handling. How to construe a rational structure out of the various secondary and tertiary aspects of nature, embodied in human awareness, but accorded no scientific status, was the project Burtt identified as "the great difficulty of contemporary cosmology."[10] The 1932 edition of *The Metaphysical Foundations* presents the challenge.

> Perhaps we need to be much more radical in the explanatory hypotheses considered than we have allowed ourselves to be heretofore. Possibly the world of external facts is much more fertile and plastic than we have ventured to suppose; it may be that all these cosmologies [identified in the historical study] and many more analyses and classifications are genuine ways of arranging what nature offers to our understanding, and that the main condition determining our selection between them is something in us rather than something in the external world. This possibility might be enormously clarified by historical studies aiming to ferret out the fundamental motives and other human factors involved in each of these characteristic analyses as it appeared and to make what headway seemed feasible at evaluating them, discovering which are of more enduring significance and why.[11]

Burtt intended to develop a scientific metaphysics which would be a method for analysis of the categories mind creates to order, organize and make sense of experience. The logical place to begin was to identify the categories and propose a method of standardized study, based on the scientific model, categorizing and standardizing nature. The metaphysical categories would standardize and categorize our actual conscious experience of nature.

The difference was that while mathematical descriptions worked for nature and supplied satisfactory standards for theory verification, mathematics did not work in metaphysics. Burtt proposed instead the historical method, with looser standards for verification implied.

In various papers, he identified space, quantity, causality, motion, logic, and universal order as mental categories of perception in addition to the category of time. All were seen as legacies of the logical, mathematical world-view. Systematically, he began to cut scientific terms asunder from their established meanings in mathematics and show them to be malleable categories in the mind, just operational, organizing principles, adopted to lend intelligibility to experience.

BURTT'S SCIENCE OF METAPHYSICS

A year after the International Congress met in Boston, Burtt published a paper in a memorial volume in honor of Isaac Newton sponsored by the Mathematical Association of London, *The Contemporary Significance of Newton's Metaphysics*. It was written in clear anticipation of a new scientific reality in the dawning and the need for a new metaphysics, or philosophy of mind, to meet it.

> *The concepts of space, time, causality, and the like, which furnished the foundations of science for Newton and Kant, have all been abandoned, at least in their traditional form, and science is still wandering in the dark in search of dependable successors.*

In the future a new scientific revolution might abandon the concept of space altogether in exchange for something more general, such as simply "quantity."

> For even though we experienced the highly unlikely revolution of finding ourselves no longer interested in measuring the extended, we might still think in mathematical terms with reference to intensive quantities.[12]

Both of these early papers, *Real versus Abstract Evolution* (1926) and *The Contemporary Significance of Newton's Metaphysics* (1927), must be seen as preliminary to the last chapter of the revised edition of *The Metaphysical Foundations* (1932). They present the groundwork for Burtt's new logic of experience, a philosophy which could have rocked the empirical foundation of modern science by demonstrating the logical consequences of two of the most promising philosophical trends of the era, the new British analytical philosophy and Einstein's relativity. Burtt seems to be saying that these two important twentieth century currents must reconcile and culminate in a new conception of human reality. This would be reality based entirely on individual experience, understood as a mental construction of interaction between *objective physical* experience and *subjective metaphysical* categories.

Burtt's doctrine of categories would mediate between constructed reality and the metaphysical presuppositions on which constructed reality rests. It would encourage an on-going awareness of the processes of mind interacting with immediate experience. Philosophy's new task would be recognized as the discipline of working out a logic of awareness of the processes of mind through which experience is ordered and interpreted. It would define new operational categories as they emerge, explain how categories are generally adopted, and call into awareness how the metaphysical categories are used to create the ever-changing parade of world-views through time and across cultures.

Burtt introduced his "science of metaphysics" in *The Contemporary Significance of Newton's Metaphysics*. He abandoned metaphysics as ontology, abandoned certain knowledge in the classic sense, and abandoned the possibility of attaining anything like it, either empirically or metaphysically. All we can know, he decided, is how knowledge changes.

> *If we take the empirical point of view as seriously as it deserves, in what sense is it possible to work out a science of the ultimate structure of the universe?* [13]

> There appears to be nothing in the universe protected from change; there is no eternal structure of reality, from whence it follows at once that there is no science of it.
>
> But alas for our attempt to rest content with this result!... For if everything changes, then in the concept of change, at least, we have an ultimate metaphysical category which can be confidently affirmed of the universe, and all the consequences involved in the meaning of change can be developed by deductive logic and also applied with assurance to reality.[14]

Developing all the consequences involved in the meaning of change by deductive logic was a very tall order, but it was one toward which Burtt was to make some progress. The first step was to ferret out the standard against which change can be measured. The only standard to which change can be compared is constancy. We have no evidence for constancy in the universe and therefore have no choice but to begin with a presupposition. We must accept the same presupposition that Newton acknowledged and begin there. The universe is orderly and intelligible. The necessity of our assumption does not make it more than that; it does not, for instance, make it True that the universe is an ordered place or that it is an intelligible place.

We must assume order and intelligibility so that we can study the metaphysical categories mind has adopted as universals derived from the interaction of mind and experience. Time, space, causality, quantity, motion and logic are superimposed on an orderly universe once such a universe is presupposed as our operating metaphysics.

The categories are not to be seen as real categories existing in nature in one to one correspondence with nature as it is in itself, but metaphysical categories derived from experiencing the world with a scientific world-view and as a consequence of the presupposition that the universe is orderly and intelligible to begin with. As a matter of common culture and education, we experience the world this way because we interpret experience through these categories. As such, the categories are relative instruments of mind operating in a particular culture at a particular time and producing a consensus of experience within a certain microcosm. But the consensus is transitory; it can and will be replaced in the future by other categories, different instruments of the mind, adopted to describe new experiences of the only real substructure of the universe which is change.

The "contemporary significance of Newton's metaphysics" is that Newton established the necessary presupposition that "appearances form an ordered, intelligible system" as a metaphysical principle. Using Newton's *Rules of Reasoning in Philosophy*, just as he had used them to make his argument in *The Metaphysical Foundations*, Burtt noted that we are compelled to assume that the universe is an ordered and intelligible system "and we must not assume anything more."[15] If scientific understanding is to rest on as few presuppositions as possible, then this is the one. "For [the category of intelligible order] to change," Burtt wrote "we should have to cease thinking entirely, and therefore our world as an experienced object would no longer exist. And we know nothing beyond [immediate] experience."[16]

According to Burtt's interpretation, Newton himself accepted this presupposition provisionally as expressed by his third and fourth rules.

> The testimony of experiment is the ultimate criterion of truth, but experiment itself is impossible unless we assume that Nature is an intelligible order, that is, that its baffling complexity can be reduced to the simplicity of law. This assumption is therefore final and absolute. Furthermore, as a methodological principle (not absolute) we have the right to presume that bodies yet to be studied have the same qualities as have been universally revealed in those already experienced, but this is subject to the proviso that further experience confirm the presumptions. [17]

Therefore the "science of metaphysics" derives directly from Newton, is predicated on Newton's world-view and Newton's rules of reasoning. The difference is that Burtt's metaphysics will be a historical and psychological study of the intermediate categories, not a mathematical one.

> Metaphysics thus becomes the science which reveals the nature and the place of the intermediate categories between the absolute category reflected by the postulate which all thinkers have to make and the categories of the other sciences. But it is essential to observe that the allocation cannot be made merely by observing the relations of implication between the categories recognized by the sciences themselves (as contemporary rationalists are prone to suppose), for apart from the functioning of a live explanatory interest and a recognition of the ultimacy of the category it involves, the hierarchy will remain very incomplete. It will tend to culminate in the highly abstract postulates of mathematics, thus neglecting the existence of non-quantitative logical relations and the real explanatory interests they reflect.
>
> Except with respect to its one absolute postulate, then, [the postulate that the world is orderly and intelligible to man] metaphysics will be a changeable and growing science like every other, but it will less rapidly change than any other, inasmuch as its function is precisely to bridge the gap between the most general categories of other branches of thought and the ultimate postulate of all thought [the necessary presupposition that the universe possesses an intelligible order and continuity which man can comprehend].[18]

The distinction Burtt had made between "real time" and the "time of science" or "abstract time" paved the way for his "science of metaphysics" because it showed that time is dependent upon Newton's metaphysical presupposition, universal order.

> Abstract time or the time of science only emerges when the expanding process of abstract time has gone far enough for the meaning of a successive continuity to be clearly seized and the habit acquired of fitting events into it.[19]

Burtt's concept of abstract time as a metaphysical category is not like Kant's category of time as an *a priori* foundation of sensibility. It is not *a priori* because our experience of abstract time depends on the our accepting causality in time relations, which, in turn, is dependent upon the presupposition that the universe is ordered. Burtt's philosophy of science depends on making universal order, including causality and time, a metaphysical presupposition. The task of general philosophy is to make it consciously so. History demonstrates that the concepts of causality and time are dynamic controllers of experience. We can take nothing for granted, nothing as *a priori*. We have only our interpretation of experience. The bottom line is that

> ... the only world which we can be interested in and think about is the world that takes form and reality in the course of our experience. [20]

In the only contrast Burtt ever made between his own concept of metaphysical categories and Kant's categories he remarked:

> If Kant had meant by the understanding that matrix of inner concerns which determine perception and guide association, in their natural growth toward generalization and unity, he would have been essentially right in holding that the understanding is the law-giver of nature, the source of its order and coherence. Only he would have added that the fundamental types of order are so selected and formed as to provide appropriate adjustment to the play of the external compellers of attention, so far as the latter normally need to be squared with in our concerns. Thus they are patterns of connections of experience imposed upon us from the outside, while being also, in their unity and totality, the product of mind.
>
> Kant was also right in holding that the uniformity of nature, the presence in the world of a single pervasive order, presupposes unity of selfhood in the experience of such an order. Only he should have noted that since this unity rests on a process which is never complete of continual growth in the generalization of needs and their integration in a more complete system, the self implied is not absolute, but is always struggling to preserve and secure itself in the further march of experience. There is no guarantee of final success in this struggle.[21]

On this understanding, the whole enterprise of *The Metaphysical Foundations*, throwing doubt on early modern science, was part and parcel of the construction of the new age, the age to which the Columbia radicals believed themselves to be the midwives. Taken in light of one recent account of the elusive constellation of concepts which go together under the umbrella of post modernity, Burtt ushered in the bulk of them by rattling the foundation of knowledge in the physical sciences and suggesting that science is a cultural affair.

> Virtually in every kind of intellectual endeavor, postmodernism tries to show that what others have regarded as a *unity*, a single, integral existence or concept, is plural. This is to some extent a reflection of structuralism, which understood cultural elements - words, meanings, experiences, human selves, societies - as constituted by relations to other elements. Since such relations are inevitably plural, the individual in question is plural as well. Everything is constituted by relations to other things, hence nothing is simple, immediate, or totally present, and no analysis of anything can be complete or final.[22]

When the reviews of the *Metaphysical Foundations* began to appear throughout the year of its publication in England, it was clear that the book had impressed its readers. Reviews showed up on both sides of the Atlantic for more than a year. There were reviews in *The Times London*, *Literary Supplement*, and in the prestigious science journal, *Nature*. Bertrand Russell wrote a review which appeared twice, once in the London periodical, the *Dial*, and once in *Nation and Athenaeum*. In the United States reviews were printed in *The Boston Transcript, The New York Tribune*, and *The Saturday Review of Literature*.

Burtt had not burst from obscurity with his 1926 paper delivered before the Six International Congress meeting at Harvard; *The Metaphysical Foundations of Modern Physical Science* had already established his reputation.

SCIENCE AS AN ETHICAL METHOD IN PURSUIT OF TRUTH: SCIENCE, ETHICS AND PHILOSOPHY TO BE DISCIPLINED BY SCIENCE METHODS

Burtt held a variety of ideas which are contradictory if typically construed. For example, he greatly admired the scientific method because of its strict verification standards. He considered himself an empiricist, but his own rigorous application of these principles never convinced him of scientific truth if that truth is understood to be one to one correspondence with the way nature is in itself. He admired science as a method of knowing and viewed the history of philosophy as a sort of scientific record of man's attempt to understand reality. In a juxtaposition of words conventionally understood as incompatible, his vision included an empirical metaphysics of experience and later, an empirical theology.

He was an idealist, with deep religious feeling, even religious passion. But he did not allow himself to believe in God because there was no convincing empirical foundation upon which to base the belief, even if that foundation was very broadly interpreted as William James had recommended.

He was a deeply ethical man and recognized that ethics demands some truth standard. Yet, he convinced himself the only constant, the only standard in the universe must be imposed there by presupposition. For him the only reality was change. He knew the importance of establishing a standard in ethics; he believed in the value of Truth and he wanted the Truth to be one, single and absolute Truth, but he could find no method, scientific, philosophical or theological, which led him to it.

Fundamentally wedded to the idea of Reality as existing in the ideal sense, but helpless to figure out how human beings will ever glimpse it, he settled at last on a definition of reality as human understanding in the broadest sense. His admiration for the scientific method and adherence to empiricism merged with his private theology and idealism into a brand of pragmatism which made seeking Truth its highest value, even while denying the possibility of ever finding it.

Like many complex thinkers, and most radical thinkers, Burtt was trying to reconcile conflicts within himself. For him, it was the fundamental Christian faith of his father, so to speak, and the liberal education he was exposed to in America which were the roots of the conflict. When he was studying at Yale the standard program was Scottish common-sense realism based on the philosophies of Locke, Berkeley and Hume. Standard in all the universities, it was especially well entrenched at Yale. Sometimes called mental philosophy, it was really a theory of physiological psychology, also referred to as "intellectual philosophy" or the science of the mind. Dealing with different mental powers, it boiled down to psychological ethics, presented as a theory of man's "active and moral faculties." In seminary schools it passed as liberal Protestantism and could be used to support religious and moral truth with arguments belonging quite easily to secular rationalism. For Burtt it was an anodyne to his father's dogmatic fundamentalist Christianity and an invitation to think for himself, which he took to mean, to think scientifically.

Theology held natural attraction for Burtt, and he might have imagined becoming a different kind of preacher than his father had been, redeeming both Christianity and his father in the process. After graduation from Yale, he was fully intending to pursue the path well worn by previous generations of New England Burtts, the Baptist ministry. His studies at Union Theological Seminary led to an article on ethics, *Present-day Tendencies in Ethical Theory*, which appeared in the highly respected journal, *International Journal of Ethics* in 1921. This article bears many traces of Burtt's educational Odyssey up until that time as well as being a good indication of the his life-long tendencies toward idealism. Most important, it demonstrates the penetrating intellectual honesty with which Burtt faced spiritual perplexities. In the discussion, Burtt, as wishful idealist, wrestles with Burtt, the disciplined philosopher. With the benefit of hind-sight the article contains some hints of his more mature thought because it foreshadows "operational relativism",[23] the doctrine of the categories and identification of fundamental presuppositions as a new sort of metaphysics and as a resolution to the quandary of ethics without standards.

Burtt was teetering between traditional religious metaphysics and relativity in moral standards. The problem he wanted to address was how far ethical judgments and standards are relative and how far absolute because "as a matter of fact we live in a world where there are both things relative and things absolute, the flowing and the fixed, the temporal and the timeless." Burtt's admiration for Hume's philosophical discipline, "keen and frank thinker that he was," and for Hume's insights into a thinker's "presuppositions" are very much in evidence in the article. Already Burtt had adopted Hume's penetrating logic as his own searchlight. Hume "was predisposed to see relativity everywhere" as is "natural to students of morals", and Hume was to become Burtt's guiding authority on to the rigorous terrain of polished skepticism.

Burtt sided with Hume on the point that the facts of history demonstrate relativity in moral standards, but he could not leave that alone and wanted to improve on it. He made the case for a more comprehensive interpretation of ethics, tied to science and logic in order to reveal something "truly structural...something that saves moral experience from the shaky hand of the relativist and gives it a certain degree of fixity." The error he identified and faulted in other thinkers is that they try to find a single comprehensive metaphysics, a single concept adopted at the expense of everything else. The materialists or realists, for example, tie everything to the immediate environment of physical nature or to social facts. The idealists are no better, tying ethical conduct to "the ultimate environment of all existence," the absolute or universal. The new realists and the pragmatists are no help, except that the pragmatists do attempt "to strike a compromise between materialism, and idealism,... relativism and universality." In Burtt's thinking synthesis is always promising. We can see the very same attitude and method applied to the various dogmatic schools in philosophy later on in Burtt's major writing. Twenty-five years later he was an unceasing advocate for not choosing, for not taking sides. His mantra would become synthesis--both/and.

The problem for ethics was that every apprehension of Reality was based on human experience which is always limited. Out of the necessity of the human condition, experiences occur in time and are therefore moving and changing. How are we to achieve some standard of unchanging reality against which to measure the accuracy of our apprehension when all we seem to be able to get hold of is changing perception?

> Are [human experiences in temporal existence] not for that reason unsuited to be a standard, since a standard is used for purposes of measurement and must be something stable?...There is, besides concrete existences, a stable structure of reality which does afford a certain objectivity quite above time--above time because it is a system of continuities of which time is one. It would seem that an ethical standard, if it be possible at all, must be something of this sort--not a moving individual existence but a type of continuity inherent in the structure of things. Why must we fall in love, either with the flux or with the eternal?...The connectedness and relatedness

CONFLICTS BETWEEN KNOWLDEGE AND BELIEF MUST BE RESOLVED BASED ON KNOWLEDGE AND CONSENSUS OF BELIEF

By the late 1920s, Burtt embraced the scientific method as fundamentally sound and the most promising route to real knowledge. On the face of it his conviction seems to collide with his theological temperament and his fundamentalist Christian upbringing, but to judge from his later writing, he thought of it as a beginning toward reconciliation between different types of knowing. Taking Newton's presupposition as a starting place, if universal human experience is the test of truth in a world where the scientific method is the adopted pattern of operation, then the existence of God must be denied. Although unsettling, there is no choice but to accept the world as we find it and work within a consistent intellectual framework. For this reason Burtt rejected William James's argument justifying mere belief and instead made scientific method into an ethical program and assigned it the social responsibility of identifying the truth. In an essay, Two *Basic Issues in the Problem of Meaning and of Truth*, he argued the case.

> Taken as a matter of plain empirical fact, without introducing any ethical postulate, is it not fair to say that on a general pragmatic basis the meaning of any concept is just what it happens to be at any given time to any individual who is using it, and that any statement is true (as James's broadest statement of the pragmatic principle indicates) to anyone who finds himself disposed to believe it. Taken in this factual way, there are inconsistent elements in meaning of the same term, and contradictory assertions are accepted as true. Some people believe in God as existent, and others are atheists. But if we erect a theory of meaning or of truth which simply justifies this individualism, as James appears to do, we can make no social progress in our thinking. We sit down, content each to believe what we happen to feel like believing, and not worrying about the fact that other people will believe something else. This attitude is the complete negation of the scientific spirit. If we ask why, the answer must be given in ethical terms. We feel that meaning and truth carry the implication of universality, that a certain social responsibility is bound up with them, that the reflective progress we desire is precisely in the direction of such responsibility, that, in short, concepts *ought* to mean the same thing to all minds, and that if any statement is to be called true it *ought* to be possible for any interested person to verify it as such. A concept may, at present, in point of fact, mean something different to you from what it does to me, but if so, we *ought* to find some way of interpreting the meaning so that it may become a common possession. You may, at present, take some assertion to be true that I take to be false, but if so, we *ought* to devise a

> technique of verification that will determine as between these conflicting claims and definitely refute one or the other of them. It is perhaps the very essence of the scientific attitude to make this ethical postulate.[28]

Universality of understanding, then, is to supplant universal experience as a standard of truth. It is ironic that Burtt is seen by contemporary writers as a religious thinker in any traditional guise. He was nothing of the sort. In *Religion in an Age of Science*, he explained the essential conflict of his life.

> The conflict [between science and religion] is...one of fundamental attitudes, of pervading ideals as to what is of the greatest value in life. The ideal of science is that of intellectual honesty and social verifiability, pursued in an atmosphere of complete tentativeness and mutual cooperation. The ideal of religion has been that of personal salvation, attained by inflexible loyalty to some revered leader, institution, or doctrine. It is this conflict of ideals that must be once for all resolved....[29]

It must be admitted, Burtt went on, that all religious ideals will crumble in the face of scientific scrutiny leaving only "one available alternative...that religion reform itself from the ground up, to the extent of becoming through and through harmonious with the spirit of science. Can this be done, and anything properly entitled religion still remain?" he wondered.[30]

Once Burtt had made the scientific method into a socially responsible method to establish truth and had equated that endeavor with social progress, he made the methods of science, minus the mathematics, his own ethic. His future course in philosophy, not apparent before the mid-1940s, was to work out a dependable method for establishing philosophical truth based on common experience and consensus about it. Historical analysis of scientific thinking and his philosophy of science would inform his philosophical method and bring him to quite a different place than the one at which the positivists had arrived. Burtt saw the positivists as dogmatic, no better balanced, nor truly scientific than religious zealots. Science, ideally practiced by open and fair minded intellects seeks a common understanding of nature based on empirical findings. It ought to become a model for agreement on philosophical matters as well. The new realists and the positivists were not operating on the scientific ideal; they were pugnacious, reveling in disagreement. They were caught up in their own preferences, their own metaphysics of method which made them unconscious of favored presuppositions. The truth standards they professed are not verifiable as we have no universal philosophy against which to measure their claims. It is better to sweep away all claims to verifiable truth until we have a truth standard aware of its own metaphysics of method. This can be achieved only through communication and consensus.

Since Burtt wanted working philosophers to model their method on the practice of working scientists, he made it axiomatic that scientific truth is a matter of social consensus. Social consensus became an ethical ideal and the best standard for truth and meaning. Getting at this truth by understanding one another, Burtt proposed that progress in philosophy is on a par with progress in science. On Burtt's view there was no philosophic truth possible because among philosophers there was no social consensus and no method yet known to achieve it. The solution was in finding the method.

> Once [the view that scientific attitude is an ethical postulate] gains plausibility, we see on the one hand that the operations that may be accepted as established meaning or truth will be definitely restricted in nature by this ethical consideration. They must be operations such as will make possible the attainment of social agreement on the meanings in question, and any operations not of this character cannot be regarded as having any relation to meaning or truth.[31]

Science is "ethical" because it is entirely "a co-operative affair."

> Why is science empirical? Because only externally observable data that can be pointed out to the senses of other people on the same terms as to our own can furnish an adequately common basis of scientific truth...Why are hypotheses and theories to be tentatively held? Because no matter how much they may appeal to their inventor, he must not assert them as true till he has discovered some way of establishing them as such to the satisfaction of other inquirers. Why is mathematics the preferred tool of deductive analysis in science? Because mathematical thinking is the one sort of mental operation on which we seem able (for whatever may be the reason) to mean the same thing at each step of the game as our neighbor does. Why is verifying experiment an external physical performance with the aid of physical tools? Because thus alone may the process be a co-operative affair, and the achievement of each stage of the verification be subject to social check. It appears to be the business of science to give meaning and truth social universality....
>
> Is it the responsibility of science only to restrict its investigations...,or does this ethical insistence apply to all our thinking, on general philosophical questions as elsewhere?[32]

Burtt proclaimed that those who adopted the scientific method as a means to social truth would be called "Co-operationists" and the "novel and important thing" about them was not so much as to proclaim again that "man is the measure of all things as they are experienced, nor to note that all thinking is inevitably affected by selective interests which vary between individuals and groups" but rather, the "novel and important thing" that the

Co-operationists would affirm is that social agreement is central to the development of any understanding of the foundation of the processes of logic.

What does this do to the search for Absolute Truth? Burtt's co-operationists would hold that "objective universality of meaning and truth becomes an ideal of logical value to be approximated rather than a factual structure to be claimed."[33]

OPERATIONAL RELATIVISM IN THREE ESSAYS: *Two Basic Issues in the Problem and Meaning of Truth* (1929) Burtt's *Review* of M.R. Cohen's Reason and Nature (1932) and *Relativity and Scientific Method in Philosophy* (1932)

Burtt's papers from the 1920s and 1930s synthesize and extend key ideas originating with *The Metaphysical Foundations*. There is even what might be seen as an intentional published release of a sequence of ideas culminating in a complete scheme for a radically new conception of knowledge. Burtt was inclined to use the word "pioneering" to describe his radical ideas and was engaged in one sort of "pioneering" endeavor or another all of his life. But the plans for a grand design, a new philosophy of mind as a science of metaphysics, operational relativism, or doctrine of categories, became delayed, postponed and finally abandoned. He settled down to an uncomfortable and undervalued recommendation for understanding, tolerance and an "all embracing philosophy of man." The methods for achieving an "all-embracing philosophy" would be historical analysis using Burtt's own rules for cooperative evaluation of conflicting ideas.

Burtt steered a middle path between Idealism, Realism and Pragmatism, stressing a method, which could be called "Operational Relativism." It is a term he used one time in a discussion of Morris Cohen's *Reason and Nature*, which was published in *The Philosophical Review* (1931). Operational Relativism is a brand of Dewey's instrumentalism that reconsiders the Pragmatists' agnosticism about Truth.[34] While Burtt admires what he calls Cohen's "orthodox realism," meaning a strict empiricism, he wants to replace some of its assumptions with the assumptions of Operational Relativism. There is no further elaboration on what these assumptions might be, but taken in concert with the other work he was doing in 1932 the assertion probably refers to Burtt's own brand of metaphysics, "a changeable and growing science like every other," a science to "bridge the gap between the most general categories of other branches of thought and the ultimate postulate of all thought," which is that the universe is ordered and intelligible.[35] Much of Burtt's criticism of Cohen is an attack on Cohen's acceptance of an ordered universe as dogmatic, factual reality rather than as a presupposition of the scientific world-view.

Burtt's science of metaphysics, or Operational Relativism, has common ground with R. G. Collingwood's reformed metaphysics. Both sought to identify the basic presuppositions at the foundation of propositions. Burtt assumed the place to begin was with the most fundamental categories of thought derived from the scientific world-view. He, like Collingwood, was interested in laying bare the category of causality as a dynamic framework upon which experience is lived in the immediate, in the mind. Unlike Collingwood, Burtt proposed no absolutely unchanging presuppositions, and unlike Collingwood, he would argue that changes in fundamental presuppositions can be logically understood. Collingwood made the changes illogical and based on "strains" which could be historically documented but not logically interpreted. Burtt held out for logical changes because only in that case could philosophers guide civilization.

While Burtt wanted to save philosophy as man's role in the universe, Collingwood wanted to save the idea of God for man. Burtt's devotion to the scientific world-view and the logic of presuppositions forced him to deny the existence of God. Collingwood, on the other hand, held out for one Absolute Presupposition, however irrationally held, and that is the Idea of Being qua Being, or if you like, God.[36]

The best and most clearly developed published version of Burtt's philosophy of science as Operational Relativism, a preliminary to his philosophy of mind, was presented in a lecture, *Relativity and Scientific Method in Philosophy*, delivered February 10, 1932 at the College of the Pacific in Stockton, California. Burtt had been invited to participate in a visiting lecturers' series held in Anderson Hall on second Tuesday evenings of each month to join "the noted philosophers of the United States and Europe" to lecture to the philosophy club at the College of the Pacific. The lecture was both a summary and a culmination of various threads in his thinking. Unfortunately, the California trip was scheduled just a few days after the Chicago newspapers carried the news that Professors Burtt, Mead and Murphy had resigned. By 10 February 1932, as far as Burtt was concerned, Chicago was behind him; he was now professor at Cornell, and so he is listed in the College of the Pacific publication of the lectures. In retrospect, the timing was disastrous. The crisis at Chicago was an interruption to his work from which there was to be no full recovery.

The College of the Pacific lecture attracted no particular attention. It was published along with other lectures in the series in a little-known volume. For understanding Burtt's thinking in relation to recent thought (Thomas

Kuhn's, for example) the lecture is key. In the 1929 paper, *Two Basic Issues In the Problem of Meaning and of Truth*, the problematic idea that scientific truth is socially constructed was simply taken as given. Now, as everyone knows, this is a matter which was to become quite troublesome to philosophers of science who criticize Thomas Kuhn's description of scientific process. In the lecture, *Relativity and Scientific Method in Philosophy*, Burtt took as axiomatic another problem for Kuhn's readers--that scientific truth is constructed from an interaction of the observer and the observed. In some ways, Burtt's thinking anticipates Kuhn's notion of lexicon structure and disciplinary matrix as the basis of teaching/learning exemplars through which scientists acquire the context in which to interpret their empirical observations. Burtt told the College of the Pacific Philosophy Club:

> Two different universes walk around under your hat and under mine. Now if they are different...this is not because one of us is simply right and the other wrong, but because we really see something different, and that our preference for different metaphysical categories or for different meanings in the same categories is a reflection of this circumstance....To determine then how each of us came to see the world as he does, to locate the conditioning factors which have made the world appear to each of us as it has, would be a vital part of the study of the meaning of the categories. By tracing the source of personal variations in meaning it would enable us to understand each other better and take a definite step toward the objectification of the terms used....It is evident that these conditioning factors would have to be largely psychological or biographical in character because the external objects are supposed to be the same; what accounts for varying conceptions of them must be found then mainly in the different individuals who harbor these conceptions.[37]
>
> What distinctive characteristics then of scientific method does this relativity situation thrust upon us? Mainly that we have to distinguish two general steps that had not previously been distinguished in the same manner. First, we have to describe as precisely as we can what it is that we observe, not in terms of realistic independence, but in terms of a set of mutual relations between observer and what is observed. Second, we develop a technique for translating what is observed into terms of other frames of reference so that we can tell what it would look like to any other observer watching it under other conditions of position and motion. This enables us to transcend the limitations of any particular observer and to build a science that is just as objective as science had been supposed to be before relativity days. But it is of the first importance to remember that this technique of translation from one frame of reference to another is not secured by establishing a privileged center of reference, but is purely a mathematical system of invariants. It locates something in a sense absolute in the mutual relations of observer and observed, and places that absolute in an objective structure transcending any special observer. But it does not re-establish a real independent of all observation, nor of factors requiring reference to foci of observation for their accurate description.

> It will be evident now without many words, I hope, what suggestion I take all this to hold for scientific method in metaphysics and the other branches of philosophy. If the real object for science is something whose determination involves reference to conditions located in an observer as well as conditions outside him, it is plausible to hold that the objects denoted by metaphysical categories also involve references to such conditions--that it is not the world apart from this or that metaphysician that is qualitative, substantial, ordered, etc., but the world in relation to him as observer. And while the conditions in the observer appealed to by the scientists are not normally psychological, being mainly modes of mass and motion, a casual glance at the metaphysical situation strongly intimates that here, along with other individualized conditions, psychological ones may very likely have to be included. Wishes and anticipations based on them are usually regarded as mental affairs, and it seems *prima facie* very likely that the difference between the appearance of the world to the unsophisticated metaphysician who sees luck and chance in it everywhere and its appearance to the scientist who sees in it a realm of inexorable order is partly due to the fact that the one wants plenty of miracles and providential dispensations and expects to see them, while the other wants to eliminate them entirely and would refuse to admit any event as such no matter how difficult it would be to interpret it as a thoroughly orderly occurrence. At least we ought not to rule out such possibilities in advance, but be ready for whatever conditions of observation might promise to throw light on the nature and significance of any metaphysical category.[38]

Burtt's purpose in taking a relativist position was different than Kuhn's, however. Kuhn seems to have been denying the power of science to get at the truth of nature because we have not got any true picture of nature against which to fit the facts of science and therefore no reliable truth standard to judge them. Burtt would agree and see no problem here because he had demonstrated as much to his own satisfaction already. He certainly understood what is now referred to as theory dependence of method. It was Burtt who first advanced this idea as metaphysics of method in *The Metaphysical Foundations*.

Burtt's relativism was different from Kuhn's; Kuhn always claimed to be a realist. Burtt was an idealist at heart and thus his Operational Relativism was instrumental only. Part of his purpose was to make the great variety of philosophical opinion scientific by which he meant that philosophers should be able to reach consensus in the manner that he imagined practicing scientists do. Not that Burtt thought that the scientists approach the truth about nature, but that they approach the best truth about their own theories about nature. A philosophical consensus would be the best truth about man and man's relationship to nature. This was to be achieved by raising metaphysical presuppositions into awareness and then subjecting them to logical analysis. Remember that Burtt praised Newton for keeping his philosophy out of his science by holding out the presupposition of universal order as contingent.

Kuhn, on the other hand, was interested in examining how it is that a group of scientists who have arrived at consensus is suddenly disrupted by a single investigator or group of investigators who see things differently and prompt a scientific revolution. Kuhn wonders how the consensus in normal science breaks up. Burtt wondered how the consensus in normal science is established. Burtt, trying to explain why it is that philosophers see things differently, traced out an analysis of the logic of working scientists similar to the one Kuhn described in *The Structure of Scientific Revolutions*. Burtt put his analysis of the logic of science on the table as an accepted model, and a nearly perfect model of what clarity and discipline in logic should be. In contrast, Kuhn has been criticized for introducing illogic into the scientific enterprise. Kuhn's historical philosophy of science is criticized for suggesting that science does not progress ever closer to the truth about nature as it is in itself. Burtt had simply assumed that science does not give any truth about nature as it is in itself anyway and this freed him from the problems Kuhn had to address.

Burtt wanted to convince philosophers to co-operate as scientists do in their endeavor to apprehend a relative, transitory kind of truth about the physical world. He did this to open an avenue for some kind of agreement about metaphysical truth, also understood on his model, to be ever changing. He identified "categories" through which the physical world is interpreted by scientific minds. These categories were to become exemplars for philosophers attempting to apprehend the processes of mind. Burtt believed that minds of individuals create specialized metaphysical categories in order to interpret and order various fields of interest.

What is interesting in light of modern-day discussions in the philosophy of science is that Burtt defined a scientific truth as one that is transitory, one that working scientists have agreed on because they have standardized their methods of observation, measurement and logic, but one which they, like Newton, recognize as predicated on presuppositions and therefore contingent on the stability of the presuppositions. For him, the kind of dynamic truth he attributed to science, was the ideal to be had in every field. According to Burtt's thinking, there could be no more dogma of any kind if we follow the open, democratic, scientific model.

> The empirical method points out when and where and how things of a designated description have been arrived at. It places before others a map of the road that has been traveled; [others] may accordingly, if they will, retravel the road to inspect the landscape for themselves. Thus the findings of one may be rectified and extended by the findings of others, with as much assurance as is humanly possible in confirmation, extension

> and rectification. The adoption of empirical, or denotative method would thus procure for philosophic reflection something of that co-operative tendency toward consensus which marks inquiry in the natural sciences. The scientific investigator convinces others not by the plausibility of his definitions and the cogency of his dialectic, but by placing before them the specified course of experiences of searchings, doings and findings in consequence of which certain things have been found. His appeal is for others to traverse a similar course, so as to see how what they find corresponds with his report....
>
> ...An empirical finding is refuted not by denial that one finds things to be thus and so, but by giving directions for a course of experience that results in finding its opposite to be the case. To convince of error as well as to lead to truth is to assist another to see and find something which he hitherto has failed to find and recognize.[39]

What was important to Burtt about scientific truth was that a scientific truth is a universal one in the respect that it can be experienced by anyone willing to follow out the scientists' method in the manner the scientists describe. Burtt thought that this was the important task that remained to be done for the other areas of thought. He had made the same point in the 1929 essay by quoting John Dewey since the essay was in Dewey's honor.

With its roots in Dewey's reconstruction of philosophy, Burtt's recommendation is grounded in the philosopher's commitment to exacting discipline and thoroughness. For Burtt this implied not only an allegiance to Dewey's logic, but also an allegiance to studied self-awareness of the processes taking place in the mind as it operates "logically." The theologian's soul-searching for moral motives became handmaiden to Burtt's intellectual search, modeled on Hume, for unidentified presuppositions. In the introduction to his anthology on the British Empiricists (1939), Burtt remarked that while both the philosopher and the man without logical training will share an interest in the same great questions, the philosopher has an intellectual responsibility that the common man does not. For Burtt intellectual responsibility meant following certain rules common to both the philosopher and the scientist. These are "insistence on the clarity in the meaning of our terms, on consistency in the affirmations made by the use of these terms, and on humble respect for all discoverable facts that are relevant to the problems we are attempting to solve...."[40] Operational Relativism adds something more to this list--the philosophers' recognition that mind is an instrument, and not necessarily a perfect one. A responsible philosopher incessantly questions and re-evaluates the processes going on in his mind. Thinking must take place in full awareness of thinking mind as a vehicle, which uses presuppositions, necessarily, to frame ideas. The important thing for the philosopher is to be aware of every aspect of how the vehicle is operating and not to let it run off wildly on automatic pilot.

In the review article for Cohen's *Reason and Nature* Burtt had argued that as long as nature is understood to include everything except awareness of mind thinking, it will appear to include as part of nature the relations and order of events as well as the events themselves. An independent and intelligible pattern in nature appears to be inherent in nature, but actually, it is in the mind observing nature. Subtract mind by being aware of mind's processes and order disappears. The enigma is that without the presupposition of order and intelligibility mind can not think at all.

Over and over again, Burtt identified one single necessary presupposition, or universal, sometimes called a "category," not Kantian and not *a priori*, but none-the-less inescapable. This was the "category of intelligible order or continuity....a category absolute and unchangeable as a character of what presents itself as the world to anybody who chooses to think. For it to change we would have to cease thinking entirely, and therefore our world as an experienced object would no longer exist. And we know nothing beyond experience."[41]

According to Burtt, thinking mind uses empirical data, gained under certain conditions of experience. Conditions modify and qualify lived experience for each individual in distinctly unique ways. Therefore, experienced data are to be regarded as conditional and the conditions must be identified when the data are put forward in communication. Some of the conditions under which data are selected from experience are fundamental to the scientific world-view. In this sense, they are cultural and specific to this historical age in the Western world.

> People, ages, races and groups have differed in the kind and degree of rationality which they have found in their world. Many find much more chaos and uncertainty than the opposite, and it may be surmised that were we not so strongly gripped by our scientific enthusiasm we should agree with them....Is it scientific to propose an answer to the problem of the rational structure in nature that ignores these circumstances? Would such an answer have any more logical virtue that an answer to the problem of the motion of the sun which ignores the fact that there are different points of view from which the motion of the sun appears differently? And surely such variations depend on the observer as well as on something outside him....May we not say that at least some of them depend on the observer's mind?...Do not [these considerations from science and scientific reality] strongly indicate that metaphysical facts are in the same case as the rest, and that therefore some of the variable conditions affecting the rational order of the world are to be located in mind?
>
> ...And is it not rather unscientific anyway to put on such haughty airs about the virtues of the perspective from which we orderly-minded folk happen to view the world? Return a moment to our inhabitant of the south pole. As he sees it the sun moves in a most simple and seemly curve. Late in September it slips above the horizon and, circling evenly day by day,

> slowly rises a fourth of a quadrant. Then it reverses its course and after six months has disappeared once more. How rational in its neat simplicity! Surely, he must conclude, those folk in other parts who see the sun gyrating in more complicated curves are sunk in childish and unaesthetic delusions. They need conversion to the purity of reason, so that they might see the sun behave in accordance with the requirements of immaculate order. Does not this analogy forcibly reveal how impossible it is to pass by disturbing facts by insinuating immaturity in the minds which find them to be facts? Just as there were natural and inevitable causes for the sun's appearing differently to other observers than it did to the south-pole tenant, and no explanation of the sun's behavior is adequate that does not account for its appearance to all of them, so there are natural and inevitable causes for the world's appearing chaotic rather than rational to observers whose interest, expectations, emotions, and established ideologies, are radically different from ours, and no explanation of reason in nature is adequate that does not account for its variations to such observers. Moreover, it must account for them impartially.[42]

Burtt argued that realists, like his old friend and teacher Morris R. Cohen from City College in New York City, want so very much for the world to be rational that they think it is important to discover all the order possible in it. As if he were engaging Cohen in discussion directly, Burtt wrote:

> But to do this by saying that reason is the substance of nature is to camouflage propaganda in the guise of metaphysical logic; it is to deny...the facts...namely that most people find much unreason in their world and delight all too readily in it....Reason a substantial factor in the world? Yes, assuredly, and so is unreason. Under certain conditions the one will seem to dominate nature, under different conditions the other....It is important to locate the discriminating conditions.[43]

In order to locate the discriminating conditions Burtt says:

> It is necessary to postulate that they, like other objects of intelligent inquiry, reveal an orderly pattern....Clearly the dependable conditions of unreason must be as rational as the conditions of reason. But it is one thing to postulate rationality in this way as a matter of intelligent method--it is another thing to insist that it exists entirely apart from those conditions which seem to be required for its appearance.[44]

While Cohen recognized contingency in the world as well as intelligibility, he steadfastly maintained, according to Burtt, that it is intelligibility that is the world's substance, "without any systematic inquiry into the conditions of either, or any frank facing of the facts...which point toward a more complex interpretation."[45]

For all his admiration of science, Burtt was unimpressed by its companion philosophy, realism, "which blandly attributes characters to the world at large and fails to envisage the need of a more empirical, specific, and exact explanation." His own scientific metaphysics called for the study of the conditions under which a "category is absent and [to] attempt the difficult task of discovering precisely the conditions under which it is present, those under which it is absent, and those under which it is present in varying forms and degrees."[46] Intellectual history, recovered and compared from culture to culture could yield the data. Any such study would have as its method the assumptions of Operational Relativism, with a temporary suspension of the presupposition of universal order for the physical world. Implied was a method of cultural study which could abandon the scientific world view in order to enter the presuppositional framework of other world-views and then return to the scientific view in order to communicate across the differences. This is exactly where Burtt's life of practicing philosophy ended up--attempting to merge the philosophies of East and West.

By 1929, Burtt had already identified time, space, quantity, causality, motion and logic as intermediate categories of the scientific world-view. His approach was to treat these terms as metaphysical categories, invented by the processes of mind and not to be understood as part of objective reality, or as a part of the universe itself.

By the late 1920s, Burtt had given up the idea of any eternal structure of reality, which man can hope to apprehend, except the concept of change. Change is the only real category we will ever understand. Operational Relativism is a method of making change understandable and useful. The consequences of change are what we experience and what our minds make into the intermediate categories. In order to conceive of these as universals in a scientific culture and put order and meaning into experience we must presuppose that the universe of our experience is ordered and intelligible. But, we must realize that it is actually we, ourselves, who make it so by thinking in terms of the intermediate categories of the scientific world-view, time, space, quantity and causality. Furthermore, we have made these categories into universals for the very purpose of getting the mental effect of order.

Our own preconditioning into the contemporary world-view causes us to select certain data from the environment and ignore other possible data. We select the sensory input which fits the categories of our world-view and construct our ordered universe. We see and experience what we expect to see and experience. Understanding this is understanding reality to the extent that such an understanding is possible for man.

FROM *THE GENESIS OF HYPOTHESIS* (1935) TO *THE HUMAN JOURNEY* (1981)

In *The Genesis of Hypothesis* Burtt advanced the argument that the scientific world-view satisfies a particular human purpose. Our constructions of time and causality have made it possible to predict and control events, giving us power over the environment to meet our physical needs. In an evolutionary sequence, Burtt sees this need as basic and natural to human progress, our natural course in evolution on the planet.[47] In 1981, he picked up the theme again in a series of lectures delivered for the University of Calcutta, published as *The Human Journey*. Burtt thinks that human needs form a hierarchy, with the need to come to terms with the physical environment taking priority. Thus, we have science with the power to predict events and control nature. Next, humans have a need to get on with their fellows in society, to co-operate and understand each other. This is the need controlling human evolution at present and this is why Burtt believed that communication across cultures and among the world's various idea sets was the "pioneering project" of the day. Last would come the "spiritual pioneers." After "the intuition of a single unified order pervading the universe, on which science and philosophy could develop is gained," people will have satisfied basic material needs and living in relative comfort will next learn how to live together amicably. Then the time will be ripe for "contact with Divine Reality."[48] This is not something new in Burtt's thinking at all. Back in 1921, Burtt had written of the "supra-real" and the "infra-real" and wrestled with human potential to square these up together in his article on ethics.

By the 1980s, Burtt had been retired for nearly twenty years. He seems to have made peace between his life as a philosopher in the scientific age and his inner reality, his search for God. Like so many of the world's great thinkers his best work grew out of the tension created between a deeply honest, singular intellect caught up in a world-view at odds with the inner man.

Burtt agrees with Dewey's concept of instrumental ideas as ideas inspired by data abstracted from individual experience, but the value placed on truth is different. For Dewey instrumental ideas are neither true nor false and it makes no sense to consider truth or falsity. Ideas are a means to an end. For Burtt, the instrumental ideas are not to be regarded as true either, but as some part of one person's hypothesis about truth. Nevertheless, Burtt values the ideas as having potential to reflect Absolute Truth. Because, for him, some semblance of Absolute Truth will be born out of consensus. Relative truth held by an individual is some part of the larger human truth and this is the Truth Burtt is after. Burtt sees this truth as a dynamic one, but leading somehow to contact with an Ultimate Reality of some kind.

The resolution is that he wants philosophy to acknowledge one truth and one reality as a consensus which is possible and can be achieved through an ideal of logical communication. Paradoxically, given the limitations of the human condition this must be a relative "philosophy of man," made as perfect and reliable as possible. Perfect here means complete in that Burtt imagined it possible to become increasingly more aware of the variety and nuances of human truth. Reliable means that terms have been clarified and mediated among a pioneering group of thinkers, the Co-operationists, seeking mutual understanding.

The desire to find one truth separates Burtt from Dewey and a great many other relativistic thinkers who dismissed a set of certitudes about universals and absolutes as a proper goal for philosophy. Operational Relativism presupposes one truth, albeit a partial and dynamic one. To arrive at it we must first dispense with notions of finding it, otherwise we will fall into the belief that it has been found and dogmatism will follow.

As a method, Operational Relativism is pervaded by both open-mindedness and persistent skepticism. Burtt could not convince himself of what he wanted to believe most of all--that it would be possible to arrive at The Truth. This is what makes him a complex thinker and also, an unconvincing one as far as later-day philosophers, with their pretensions to certainty, are concerned. Again and again he came back to examine human reality as a construct of human experience, culturally and socially conditioned, and to an analysis of scientific facts, as not fixed but colored by the individual "valuations," sometimes little more than personal predilections. When Operational Relativism is the reigning philosophy, the resulting perception of reality is not Reality Proper but human experience as human reality and all we can ever know. For man some reflection of Reality proper will be apprehended when there is complete awareness of presuppositions and how they change in the minds of human beings through time has been logically explained. The purpose of intelligent living is to guide the logical transition of presuppositions from age to age.

This is Burtt, the honest intellectual. But for Burtt, the inner man, the fact that the human condition permits only the understanding of change must not exclude the actual existence of something more, a substructure of reality that is beyond man. He wanted to see different cultures, conflicting religions, and contentious factions in philosophy reconcile. To succeed he would need to reconcile between his own Janus faces, tough-minded philosopher and religious seeker. The compromise he made between these two conflicting parts of himself became his new philosophical method for clarifying terms and reaching consensus and his belief in a world synthesis as "best truth." He called it an all-embracing philosophy of man. A perfected philosophy of man will be able to recognize Divine Reality as one part of human expression. It is not intellectual softness, but intellectual compassion, hard negotiation and compromise that Burtt was recommending. Success in reaching understanding would be the mark of human maturity, intellectual, social and spiritual.[49]

Burtt proposed that maturity in a single individual or in world civilization is indexed to the degree of self-awareness attained by anyone regarding his own thought processes and the degree to which this self-awareness expands to include all the various realities of human experience the world over. Mature philosophers become masters of their own thinking minds and eventually will master the patterns and complexities of many other minds. Over the course of his life, Burtt was to dance around and around the idea of teasing the individual's lonely experience of reality, colored by his special interest valuations and his unavoidable emotions, away from Reality proper.

Burtt was right in the middle of developing a doctrine of categories to support his philosophical view during the crisis at the University of Chicago, his resignation and move to Cornell. After the lecture at the College of the Pacific he was caught up in the time consuming activity and pressure of beginning again and teaching in a brand new field, the history of religion. Burtt did not publish, nor did he finish what would have been his own original philosophy of mind. It was based on the presuppositions of Operational Relativism, a doctrine of categories derived from the scientific world-view and offered as exemplars for understanding other world-views, including a spiritual one.

The disruption in his career at Chicago was followed by marital troubles in the late 1930s, marital separation in the early 1940s, and an intense, emotionally expensive, period of psychoanalysis through the mid-1940s. In a final blow, Arthur Murphy arrived at Cornell in 1945 where he was installed as the new chairman of the Philosophy Department. Murphy did not have much use for historical philosophy and before long all the older Hegelian-type philosophers were gone. The analytic wave had begun.

It seems likely that Burtt had already put his new philosophy of mind aside anyway by the time Murphy came on the scene. Family troubles are only part of the story. His interest was shifting rapidly away from the philosophy of science and toward questions in history and philosophy of religion. This assessment is strengthened by comments Burtt made to his student, Stuart Brown, around this time. Burtt told Brown that he had been wrong, after all. According to Brown, Burtt gave up being a "hard-nosed logician" then and turned to religious interests.

His success with books on the history of religion and his fascination with psychoanalysis, Eastern religion, and Eastern philosophy were to become the corner stones of his reputation. His early work with *The Metaphysical Foundations* faded in the minds of his late contemporaries as they came to identify him with the philosophy of religion and to respect him for his spirituality, rather than his original contributions to the philosophy of science.

Examination of the lesser-known papers, dating from the 1920s through the early 1930s, illustrates quite clearly the direction Burtt's new philosophy of mind was taking. It bears similarity to parts of Thomas Kuhn's thought, especially some of the more controversial aspects. Perhaps it is not so remarkable that Kuhn's ideas have caught the attention of historians and philosophers of science some thirty-five years after Burtt first explored them. If chance favors the prepared mind, it was Burtt's *Metaphysical Foundations* (steadily in print since 1924) and the work of Alexandre Koyré, built on Burtt's shoulders, which prepared Kuhn's mind and the minds of those who became Kuhn's enthralled followers as well as his ardent critics. By tracing American intellectual history back to the young radicals around Columbia University in the 1920s we can see how the seeds for post modernity were sown by university teachers in America in the middle of the nineteenth century, how they germinated in people like F.J.E. Woodbridge, James Harvey Robinson, Charles Beard and John Dewey; how they took root among the young radicals and sprouted best of all in E.A. Burtt, whose *Metaphysical Foundations* stands at the beginning of an important thread in the story of twentieth century intellectual history.

Burtt staked-out a claim on the future of twentieth-century philosophy in *The Metaphysical Foundations* by calling attention to the need for a new philosophy of mind and indicating just how it might develop. It is clear that he had intended to work on this problem himself. His aim was to overcome Cartesian dualism with a new philosophy of mind, which would save a place for the "soul" of man, as the mind of man, thereby bringing man "back into the world of nature as part of her domain." As he says in *The Metaphysical Foundations*:

> Someone might have carried over the teleological method of explanation, discredited now in physics, to the modern analysis of the human mind; nature might have been abandoned to mathematical atomism while the other side of dualism might have been accounted for mainly in terms of purpose or use. This did not happen...[50]
>
> It is of the first importance that [the historical study of the origins of modern science undertaken in *The Metaphysical Foundations*] be supplemented by a sound insight into the major factors which have conditioned the rise both of the medieval metaphysic and of its mathematico-mechanical successor which is now seen by all thinkers to demand thorough critical overhauling. Without such insight the new

metaphysic, when it arrives, will be but the objectification of the mood of an age, perhaps fitful and temporary, rather than the reasoned expression of the intellectual insight of all ages. Unless we can approximate more closely than has yet been done this generalized interpretation, the new cosmology will hardly be worth the effort required for its construction.[51]

CHAPTER 8

FROM *The Genesis of Hypothesis* (1935) TO *The Metaphysics of Empirical Theology* (1987?)

The Genesis of Hypothesis, AN UNPUBLISHED MANUSCRIPT

The Genesis of Hypothesis is a draft of an original and exciting philosophical work. Typographical errors and word omissions, only half corrected by what must have been a cursory proof reading and editing session, betray the urgency and haste with which it was composed. It represents two decades worth of thought and is the culmination of ideas first presented in *The Metaphysical Foundations*. It is the last of Burtt's attempts to hammer out a grand scheme, and it is, conceivably, the last of his distinctly original thinking. *Genesis of Hypothesis* is Burtt's attempt to make the reality of lived experience scientific and concrete by explaining it through "metaphysical" categories of perception, which he wanted to see as having evolved biologically according to Darwinian principles.

The manuscript was found in 1994, piled haphazardly among articles by various authors and on first appearance might have been mistaken for a paper submitted to Burtt by a student or colleague for his professional opinion. Certainly it had not seemed to merit any favored treatment over the years, and no part of it appears anywhere in the published work. Still, Burtt had hung on to it. There was just one type written copy, obviously the original, with some handwritten corrections.

It is this manuscript that represents the watershed in Burtt's thinking and it marks the last of his effort to develop a new philosophy of mind. Of course, there was much publishing and writing to follow, but the later work is not ground breaking. It tends toward summary, exegesis of others' thinking, attempted synthesis and conciliation. *Genesis of Hypothesis* belongs to Burtt's glory days as a startlingly fresh philosopher. In it he attempts the demonstration that time, space, and causality are categories of human construction, without any necessary correspondence with the world of nature. Burtt's new metaphysics of categories was intended to qualify the empirical foundation of science as it had been construed since Newton.

The document consists of seventy-nine pages, beginning with a page numbered 39 and running through page 118. The curiously numbered "first" page is headed, section I. What happened to the first thirty-eight pages is a mystery, since the work seems complete without them. It is divided into six sections in all and laced throughout with references to "this volume," this "pioneering project," etc. From time to time Burtt alludes to future parts of the project, clearly indicating that *The Genesis of Hypothesis* was intended to be the inauguration of a major work in philosophy, a synthesis of all he had been working on since *The Metaphysical Foundations*. Without a doubt, Burtt scaled back the plan and then abandoned the work, unfinished.

The *Genesis of Hypothesis* has significance for Burtt's philosophy in general because it ties together many different strands in his thought from the late 1920s and early 1930s right on through his retirement and old age. In this manuscript, Burtt was attempting a classic sort of proof on the model of Hume or Spinoza. The evidence for the proof is based on Burtt's own mental experiences, organizing and generalizing them with the whole of human experience. It is the attempt to fulfill the early promise to develop a doctrine of categories for the universals adopted in the scientific world-view, time, space, causality, etc.

Section I begins with the question, what is it we reason about? We reason about that which puzzles us. In order for us to frame the puzzle, we must have a minimum amount of "data." The "data" must be sufficient to permit classification and descriptive analysis. "Data" come from empirical observation and also from imagination. By imagination Burtt seems to mean that part of reasoning which allows us to form concepts because he uses as his example our perception of space and matter as universals, "conceptually realized by imaginative extension as the whole of matter or the infinity of space." Data, which permit reason, are simply that which is taken for granted.

Section II begins with the question, how does an hypothesis get born? Relying on the language and concepts of the group of psychologists known as "associationists," Burtt says that "the datum (or some part of the data discriminated by analysis) calls forth an idea of some object or event which has been associated with it in the reasoner's past experience, and this idea either becomes his hypothesis or supplies a portion of the material out of which an hypothesis is constructed" (GH, 42).[1] The factors which determine this process are the factors of "associative recall" and these are determined by the so-called "laws of association." These have been identified by psychologists and accepted by Burtt, who wants to consider how different entities become connected with each other in experience and how such connections, once established, function in associative recall. He asks whether the kind of associative recall employed in the genesis of hypothesis is free or controlled association, and if controlled, then how? He decides that it is a form of controlled association and that is it controlled by purposeful attention. It is controlled by the purpose of explaining the data. "When engaged in explaining puzzling circumstances only ideas relevant to the explanation are likely to occur" (GH, 43).

Moreover, the hypothesis will be guided by "our assumptions as to what constitutes a sound explanation" and this is variable depending upon the cultural context in which the reasoner is operating. "Our associations in reasoning depend on our assumptions about the nature of explanation, and those in turn are relative to the social background from which we inherit the fundamental postulates underlying thinking and to the extent of our disciplined participation in that social milieu" (GH, 45).

Those familiar with Thomas Kuhn's philosophy of science will recognize a similarity between Burtt's point and Kuhn's explanation of normal science. In normal science a community of experts has been trained within a particular "disciplinary matrix" and has worked out science practice puzzles or "exemplars" posed by their teachers. Working within the same "social context" scientists all see the puzzles in the same way and can come to agreement on the most tenable hypothesis to explain the phenomena in the puzzle. Burtt is saying that the whole of scientific mankind will tend to see the world in the same way that the context from which we operate is mutual, but not necessarily actual.

He says that as an hypothesis is tested for verification there are at first a large number of ideas which have some relation to the datum in associative recall. "Just because an idea comes to mind, that does not warrant the idea as adequate to explanation." Some are irrelevant and thus eliminated. At first there is "partial indeterminacy" of controlled association and some disciplined free association occurs. This free association is limited within the "sphere of relevance" and that relevance is "limited by the degree to which the established intellectual heritage of our social group is accepted and applied."

The section ends with a summary of the laws of free association. They are:

> 1. Contiguity and similarity. That is we associate like with like and we associate those things which occur together in time. Burtt says the law of similarity is presupposed by the law of contiguity.
>
> 2. Frequency, recency, intensity, duration, primacy, and context control association. This means that we choose to associate first those things which make the strongest impact upon us. The second law presupposes both contiguity and similarity.

> 3. The quest for explanation garners our purposeful attention and acts on the first two laws to control association. In this way purpose, need or human concern plays the most significant role in selecting what will be recalled and associated.

The third section poses the question "Why do two objects or events ever become contiguous in experience?" Why, in other words, when we perceive one thing do we pass our attention from it to the next. How do we choose the next object of perception? Here Burtt chooses his own answer from psychology, not physics. He declares that the passing of attention is not determined by real structures. It is determined, instead, by something in us. If our associations were directed by real structures and in correspondence with the world as it is in itself, then "we would not need to build up our scientific knowledge. We would already think in accordance with it" (GH, 56).

Burtt's proof rests on his own experiences and homely examples from which he concludes that the factors controlling attention rest both within us and in external nature, since we may become momentarily distracted by a loud noise or pleasant sight. But ultimately mature reasoning is controlled by the present purpose or interest as the most significant factor in attention. Such purpose comes from inside.

Section IV further explores the idea of the external versus internal controlling factors. Burtt acknowledges that there can be distraction from within as well as from without. Factors that control attention do come from outside as well as within. Examples of external controls include unusual size, intensity, duration, striking color, motion or change in a perceived object. Internal factors might be expectation, mood, emotion as well as purpose. Even bodily factors can play a role in compelling attention, pain for example.

But in spite of external diversion, the conclusion stands. "When presented a datum, associative recall takes place. The connection is governed by laws of association and by laws of control of attention." The latter are internal, external or bodily factors. Among these, bodily factors figure hardly at all. Both internal and external factors might seem to be equal in value, but for Burtt, who bases his conclusion on his own experience, the internal factors are much more significant by a factor of twenty to one.

Summing up his self-analysis Burtt says, "I imposed a pattern of my own on the things to which I attended; in passing from one another I followed some dramatic, logical, esthetic, or causal net-work of relationships, that is, relationships whose structure was determined by their harmony with some set in my mind" (GH, 68). Internal factors hold attention on an object or direct attention in a definite series. Were this not so individuals would be forever distracted and never able to co-operate in a group to formulate hypotheses as scientists do. As the attention of this or that individual wanders, group attention would fragment, unable to share objects of attention as myriad external stimuli engaged them individually. Instead, we make "purposeful use" of available external stimuli by concentrating our attention on them in a pattern, and we do this as we are encouraged by culture or by the group.

One way the group conditions our attention is by setting standards about what qualifies as worthy of our attention and what qualifies as reasonable explanation of puzzles. "There is selection by the controlling set of our mind which is...constituted by our assumptions about the nature of explanation and their concrete expression in the particular question we have raised. In short, only associations congruous with [the] quest for explanation are likely to be recalled; others are not resurrected at all" (GH, 73). Mature reasoners learn to tune-out anything and everything not relevant or distracting to the attention. Experience teaches what has and what has not got relevance and that which has not got relevance does not present itself in the stream of associative recall, even the free form or stream of associative recall in which scientists engage to generate hypotheses.

Although emotion might compete to play a role, as we become more rational we discount it as fully as we possibly can, "although we never reach the point where ideas due ultimately to an emotional connection no longer pop into our minds at all" (GH, 75). The scientific ideal is objective, rational, stripped of emotion, but nearly impossible to achieve. Compared to external factors, emotion might be considered important because although we try to exclude it we are never successful in complete escape.

In Burtt's later philosophy, an emphasis on "valuation," the preferences of a thinker, eclipsed what Burtt here identifies as simply "purpose." The purpose is to solve the puzzle, but preferences of a subjective nature intervene to redirect attention and push it toward a favorite solution. Again and again, he made the point that the most constant internal control is purpose. "To be dominated by a purpose means to be engaged in the selection of the means for the realization of desired ends." It is the tension created by internal factors, wish, desire, purpose, plan, expectancy, attitude, mood, temperament, and emotion that operates in us "to select perceptions and guide associations." At last, Burtt concludes that reasons for data selection distill into an over-riding general "concern" or "interest" (GH, 75). "When, in the presence of a datum, hypotheses are suggested by associative recall, the original source of the connection now functioning between hypothesis and datum in all but a minute and negligible fraction of cases is to be found in the controlling force of some concern or interest of ours" (GH, 80).

The purpose intended drives data selection, influences association from recall and decides the ultimate forms the data in hypothesis formation will take. Burtt is clear that the desire for "Truth" does not figure into data selection or hypothesis formation. Only later in checking the hypothesis against the cultural context are standards of truth applied. The standards might be methodological, cultural or personal, and they correspond entirely with the categories Burtt had identified in the Metaphysical Foundations as unconscious metaphysics.

Section V takes up an analysis of the internal factors and ends with an appeal to Gestalt psychology to explain the instant of illumination when an hypothesis is born. Like Kuhn, who also appealed to Gestalt theory for explanation of the sudden shift in scientific puzzle solving, Burtt sees hypothesis as an imaginative synthesis, "a case of insight accompanied by a feeling of release and confidence. The synthesis thus produced has it own novel quality and organization, the latter being controlled, of course, by the particular question at hand, and by the underlying assumptions as to the nature of sound explanation" (GH, 84). Such synthesis is known not only from science, but also in artistic creation and philosophical speculation.

The section ends with this summary of how hypotheses come to mind:

> 1. Data are classified and analyzed sufficiently to render associative recall possible and give it direction.
>
> 2. Awareness of similarity and dissimilarity is the determining factor involved in the classification of data. Similarity and dissimilarity are also necessary to the process of associative recollection.
>
> 3. The laws of association condition the present availability of explanatory ideas.
>
> 4. The factors controlling the passage of attention are:
>
> > a. The almost negligible role of external and bodily factors.
> > b. Human needs or concerns, which are ultimately responsible for our selecting the content of our hypotheses.
>
> 5. Power of organizing particular ideas into complex hypotheses, which is the imaginative power of mind to perform synthesis.

It hardly needs to be pointed out how similar Burtt's thinking here is to Thomas Kuhn's description of scientific puzzle solving. Kuhn has said that a practicing community of scientists decides on what qualifies as a scientific problem. The problem solvers operate from the same lexicon structure and disciplinary matrix of puzzle exemplars, encountered in their formal scientific educations. New puzzles arise when there is a bad fit between the training exemplars and the phenomena of the puzzle. Someone will suddenly experience the similarity/dissimilarity tension in a new light. Some synthesis of ideas will occur and a critical mass of practicing experts will be converted. For Burtt the human needs or concerns that figure in data selection and theory formation are the needs of scientists to solve the puzzle and to solve it within a context which can be understood, and that is one which qualifies under the standards of a good solution set by the community of scientists.

While Kuhn did not tread heavily on established notions of objective reality in *The Structure of Scientific Revolution*, his readers and critics were quick to do this for him. Some startling implications were there, and, as is well known, Kuhn was pursued to clarify his position for the rest of his life. Burtt was far bolder and left no doubt as to his own position.

In the first paragraph of section VI Burtt comes right to the point. The laws of nature are none other than the product of mind's internal controls on recall and association, expressed mathematically in order that they should be consistently verifiable by those initiated into scientific culture, and formulated for the purpose of prediction of and control over the physical environment. The basis of the conclusion is presented as a proof, which Burtt claims is "succinct though not sufficient." It is worthwhile to quote Burtt's own summation and limited "proof."

> Let me state at once the major conclusions which, in my judgment the evidence clearly [shows] when we develop the philosophical implications of the above analysis and in doing so elaborate the position [that it is internal factors which determine sequences of perception]. They are; first, that the laws of nature, i.e., statements of types of orderly connection between events in our world depend on the laws of association; and second, that the categorical patterns or ultimate modes of order assumed in these laws rest so-far-as their unity and continuity is concerned, on pervasive human concerns. (GH, 87)

Our understanding of nature is based upon our need to see the world in a certain way. We develop categories or mental generalizations through which we experience the world and make it "real" to us. The categories Burtt had identified in his previously published papers are: time, space, causality, motion, quantity and logic. His argument is that these universals are based on purposeful selections of data from experience, ordered not by nature itself, according to any natural laws discoverable by us, but by mind processing stimuli. We invent categories of experience, and later interpret our experience through those same categories as if the categories were part of nature rather than as they should more properly be known and experienced as categories of human abstraction.

The remainder of the proof runs as follows:

> Laws of nature are formulae of regular connection between events which we hold to be true of those events; now all true formulations must have first

> occurred to us as, possibly true, every such formulated connection occurs to the mind in accordance with the laws of association, and therefore laws of nature depend on (i.e. are only conceived through the operation of) the laws of association. So much for the first proof. Now for the second. The laws of association only resuscitate, in the present, connections originally established some time in the past under the laws of shifts of attention; the laws, however, which are exemplified in all but a small fraction of shifts of attention are laws of control by the internal factors, and the common feature in all the internal factors is human need or concern; therefore, the general modes of connectivity assumed in the laws of nature are in their major features, at least, expressions of human concern....It is evident that the last steps of this demonstration involve the position that the relations appealed to in scientific explanation are properly due to internal factors. When that position has been fully established the proof will gain cogency (GH, 87-88).

Burtt argued that taking sense data as objectively real, without qualification, is a virtue in philosophy as it is in science, because in doing so we are following "common sense." This is the only starting place we have and all we can ever know of nature directly. So, too, it is all we can know of mind, directly.

> *Philosophy should, in my judgment, always commence its analysis with the tried and tested assumptions of common sense, only modifying them under the inescapable compulsion of hard facts. Since we seem in danger of being forced to conclusions far removed from the beliefs of common sense, it seems to me appropriate that we should begin with every concession to common sense that we justifiably may. (GH, 88)*

There are two special "fundamental features of human experience" that Burtt regards as "consonant with this spontaneous realism." One is the qualitative content of what we perceive, especially that content which distinguishes one object from another such as degrees of intensity of color, size and so on. The other is the "relation between two perceptions" when attention shifts from one object to another. By pressing the limits of immediate experience to include all of which mind is immediately aware, these "subjective factors" become empirical by common sense.

Using common sense to analyze the sequence of attention and consequently the sequence of perception and the experienced relations between perceptions Burtt's argument is reminiscent of his 1926 paper, *Real versus Abstract Evolution*. During the proofreading he changed the wording in the manuscript to bring it in line with time as a category by striking out the word "phenomena" and penning in the word "events" in the passage quoted above, indicating a time component, the importance of change to perception and the process of reordering experience into a seamless unity. We actually experience events singly in jerky fragments, but we make single, individually occurring events into some relational whole. This process follows laws of association, determined by some purpose, some end, some "teleological connection" between the experiences as external stimuli and a human concern, a "desired end connected to imagined means." It is purpose or concern that holds the events together or inserts events and ideas from associative recall. Our stream of continuous attention is controlled by purpose and it is purpose held in the mind as intent that gives us a stream of perceptions which hold together. Burtt wants to argue that the world in itself is not uniform or orderly. It is not a cosmos, but a chaos. Our understanding of the world through our acceptance of the laws of nature, makes the world orderly to us because it primes our associative recall and triggers the first presupposition of our culture, that the universe is orderly and intelligible.

What, after all, Burtt asks, is the ultimate relation between the laws of nature and the laws of association of ideas? Either the laws of nature constitute a selection within the field of the laws of association, as the idealist believes, or there is an actual correspondence between the laws of nature and the laws of human association where the laws of association cause us to "unite entities with their real relations."

Burtt's position rests on discounting the second possibility. He argues that the relation is apparent only, in spite of the fact that our explanations are occasionally accurate. The horns of the dilemma are that it is only because we know the laws of nature as part of our context for association that we continue to experience nature in accordance with those laws. We are so conditioned by the mental context in which we experience stimuli that we are not able to experience the stimuli as they are in themselves. The principles of similarity, universality, contiguity, and causal connection are dependent on associations in experience. The degree of confidence we have in any law of nature depends upon the regularity and frequency of past experience of the relation expressed by the law and the absence of contrary cases, and our pre-conditioning prevents us from admitting contrary cases. Burtt puts it this way: Context is a factor in association. Associations which harmonize with other associations already present in the mind are favored. Laws are accepted which fit the established structure already described by those very same laws.[2]

Now, why are the laws of nature connected to the laws of association? Burtt gives four possible reasons. First, the two groups might be logically dependent upon some other more ultimate principle or group of principles. This idea is dismissed on the grounds that it leads toward theism, God as intelligent creator. This was the possibility raised by Spinoza, implying that in Divine thought the laws of nature depend on the laws of association, making God and nature identical. Moreover, we must then think God's thoughts to conceptualize truly about His creation. Burtt discounts this out of hand as offering nothing valuable for human scientific or philosophical understanding. It is old ground.

A second possibility is that the laws of association depend on the laws of nature. The laws of nature are the laws of association, similarity, contiguity, frequency, and context and they arise in us from the molding influence of nature's structure on our ways of thinking. Since not all associations lead to laws of nature, this, too, is discounted. If it were true we would never fall into error or superstition. If the laws of nature correspond with the laws of association, our very first hypothesis would always be correct. Experience shows that there are various conceptual explanations and many errors. "Were human associations everywhere dictated by laws of nature which are objective in the sense of being valid for all minds--and we assume the real laws of nature are thus valid--this would certainly appear a quite inexplicable situation. We should expect in that case that knowledge about the world should among all peoples and at all times, however remotely separated from each other, exhibit common results as well as common principles and methods of interpretation..." (GH, 95).

The third possibility is that the two groups of laws are interdependent, "some of the laws of nature controlling the course of association, others being themselves controlled by principles of association." We associate experiences in accordance with temporal and spatial juxtapositions and we do this quite naturally from birth. If the third possibility were accurate, then our perceptions of continuity in space and time would control reason by controlling associations. In other words, there is a real continuity in space and one in time and our thinking is interdependent with those realities. However, for Burtt, objections are easily raised if we accept direct experience as it comes to us, in the order it comes to us. In a case made similarly for his lecture in 1926, *Real versus Abstract Evolution*, Burtt notes that interruptions in spatial and temporal connections by associations from recall show there is no direct, linear connection in actual experience to concepts ultimately formed. The order of discovery does not determine the order of meaning. "The contiguity of association does not seem to be patterned after the nature of time and space as science has come to conceive them." If we really did experience time and space as they are given in direct experience, then experience would be irregular and jerky, rather than continuous, as we perceive it.

To make the case that emotional factors play no small role in directing attention Burtt turns to an examination of causal relations. Appealing once again to the psychologists who, he says, gave up the principle of causality as a universal principle of association because other factors such as recency, duration, sensory or emotional intensity have been shown to determine the emergence of ideas rather than primacy and frequency, directly related to causality, Burtt argues that formerly favored factors are likely to produce ideas in association which are not causally relevant at all. "An event may have been recently associated with another in our experience, riveted to it by a strong emotion, without being its cause or effect. Any time, in thinking of any event, our associations may be dominated by these factors. Association is thus not dependent on the causal structure of the world (GH, 97-98).

In sum, "There is no feature of what we take to be the real structure of nature which by itself is able to determine the course of our associations." And also, "There is no law of association which is controlled by the real order of nature" (GH, 98).

The fourth and final possibility is: "What we take to be the laws of nature universally and necessarily depend on the laws of association; they are a selection from connections rendered available by the latter....On this basis the relativity of explanations of nature to differences of age in the individual and differences of epoch and established habit in the group are precisely what we should expect. Each individual must learn to respect in his own reasoning what his elders call the laws of nature, because his first unschooled associations naturally follow various principles some of which are inconsistent with the kind of order those laws assume. Each person needs laboriously to acquire a technique of control over his mental sequences so that he can discriminate on demand between a connection which promises dependable guidance in dealing with the entities connected and one which insinuates itself with no such promise" (GH, 98-99).

In coming to terms with the scientific world view Burtt asserts that the general patterns of order which we assume--time, space, causality, correlation-- are generalizations of the connections selected by subordinating the influence of certain association controls to others. Time and space, for instance, are established by the law of contiguity as we eliminate "jerky" contiguities and "confine ourselves steadily to the regular steps of next-to-next." Likewise, causality and correlation are generalizations of the connection established between events when we "form the habit of checking" associations favored by the factors recency, intensity, duration, etc. Consider what kind of world we would live in if we favored different laws of association. If, for example, we favored not connections by similarity, but by difference, if instead of contiguity, we favored separation, if instead of frequency, infrequency, then we would live in alternative worlds of human reality. Burtt "clinches" his argument with this: "I submit that the structure of the world as we now conceive it would utterly disappear."

At last it is time for the question why is it that we have come to favor certain laws of association and to build the structure of the world as we know it. Why have we generalized our experience into categories to produce the patterns of order in just the way we have? The answer Burtt gives in *The Genesis of Hypothesis* is one that he was to pursue in the philosophy of his later years. It is that the selections we have made are "due to the pressure of certain insistent human needs."[3] Foremost among them is the need to make sense of experience, followed by the need to cooperate in society, to predict events, and to plan for the future. The key to understanding the processes involved in changing human conceptions of the world, the development of science for example, is to recognize that our needs as a species grow and our conception of the world in relation to us evolves. Our needs form a hierarchy and base of purpose from which our interests broaden. The laws of nature are simply understood as an expression of our need to live in an orderly universe.[4]

In his lecture at the College of the Pacific, the objectivity of mathematics is compared to the subjectivity of metaphysical concerns, which, for Burtt, is how we interpret experience. The degree to which our objectivity has developed has been proportional to our need for it. In most practical matters we have found a sufficient degree of objectivity to communicate through language and cooperate adequately, but mathematics has been especially helpful in many tasks and so has been adopted and made most objective of all. Mathematics has been a simple evolutionary development.

> Why is our arithmetical and geometrical reasoning objective in the remarkable degree that is? Because, in order to cooperate with each other at all in dealing with any objects we need at least to be able to count them in the same way and to locate them in the same directions and spatial relations. A group of savages could not even engage in a hunt together unless they had essentially the same arithmetic and geometry. In short, the forms of social cooperation that are required to meet the elementary practical needs of human life compel us to form our concepts of number and of spatial structure in the same way, to mean the same thing by them, to build them into an identical system of relations. Hence we keep pace with each other in our mathematical thinking and objective demonstrability is possible. Why is mathematical physics possible--the quantitative definition permitting mathematical elaboration of such concepts as time, motion, energy, mass, and their derivatives? Because while it is not so absolutely imperative for the rudiments of social existence that we have identical meanings for these concepts to expect them to be quite objective from the start of conscious experience, yet they are so closely related to common needs of social adjustment that as soon as it occurred to scientific minds to define them exactly and ally them directly with the concepts of space and number, they found the minds of the rest of us already prepared to follow them in this step, so great advantages were immediately obvious. With respect to time and motion, indeed, the beginnings of the transformation had been already achieved under the same influences that had created the objective notion of space, due to our constant need to

> anticipate the future in some definite way and to predict how any ongoing motion is to take shape in its subsequent phases.[5]

The key question is "Why cannot metaphysical concepts be made objective in the same way?" Because "an objective metaphysics is from the biological standpoint a luxury rather than a necessity...In terms of the fundamental practical needs of living creatures there is no necessity that we should all have the same metaphysics or mean the same thing by our metaphysical concepts."[6] Burtt's purpose in looking at this question was to examine the possibility of achieving the luxury of a scientific metaphysics. If it could be determined just how we categorize experience to give us mathematics and science, then we might use that insight to consciously develop the luxury of an objective metaphysics.

In 1931, this is where Burtt's thought was going. By 1935, when *Genesis of Hypothesis* was written, other pressures just on the horizon would draw his attention and energy away from the project. In *Genesis of Hypothesis*, he pursued the evolution of the categories of experience which became the categories of the scientific world-view.

It is the scientific world-view which brings order out of disorder. Science is an abstraction from particulars, from disorganized singulars. The particulars by themselves would be chaos, but by generalizing them into time, space, causality, quantity, motion, logic, we produce a cosmos and give ourselves a place in it. From this place, we experience our needs. Looking backward to primitive man and his needs we can see how the common-sense notions evolved. Because we need food, for example, we must generalize our experience into space in order to plan to move through it to the food. Fundamental survival needs orchestrated the evolution of human perception by directing attention driven by the intent of purposeful action. Every perception in the sequence is related to the purpose, fulfilling the need for food in this example. External stimulation, which makes up our experience of the physical world gives us perceptions of motion, direction, duration, size, etc. The categories derive from those perceptions. The experience of "moving" through "space" gives a series of perceptions which are recalled again and again from memory and associated according to the generalized category, space. This gives us our experience of space, until we learn something different.

> *Space is the pattern which results from this mutual accommodation between the unified continuity of distance and direction due to the generalization of need and the insistent but irregular and accidental aggregate of connections of perception due to the external controls. (GH, 109)*

The same can be said for time, quantity, causality or any of the other categories. On time, for instance:

> *Duration and motion evidently are characters which require the category of time if they are to be successfully fitted into the pattern of our generalized purposes, and intensity involves time in its measurement.*

On quantity and motion:

> Size and motion require the category of space, likewise striking color since it has to occupy a stretch of extension.

In fact, Burtt argues, all the external factors of attention control are spatio-temporal affairs and it is their spatio-temporal nature to which adjustment of experience is imperative. It is the pressure to adjust to external stimuli that enters "vitally into the form taken by [the] general categories of reality, even though those factors could not possibly account for the unity and infinity which the categories [seem] to exhibit. Now, it is true that the categories such as space seem to be objectively real and totally lacking in any teleological character." But this is due to the "accommodation of its structure to the relations of perceptions externally imposed." It is a kind of unconscious metaphysics that space seems to be a "neutral entity, material both for the continuous prosecution of our plans and for prompt accommodation in any situation to externally imposed disturbances to those plans." In other words, its character as "externally real" derives from constant external stimulation threatening to disrupt the execution of plans and its character as a "unified principle of order" derives from the overarching sense of purpose held within the mind.

Space emerges as a generalized need and we naturally find it imperative to distinguish on short notice relations which readily fit into it. Having made space a part of the real world, we regard things which fit into space as "part of the real world and their universal types become categories of nature; [dreams and illusions, which do not fit into space] we consign to the unreal realm of fantasy." The distinction between real and fantasy is metaphysical habit which equips us to live in an ordered cosmos shared by others.

Burtt's doctrine of categories of the scientific world-view makes time, space, causality, and the rest, truly metaphysical as intermediates between universal order and our interpretation of experience. The categories are instruments of interpretation adopted by the mind, but they are not necessarily permanent nor uniform across cultures. This is surely an extension of Burtt's science of metaphysics introduced in the 1927 article, *The Contemporary*

Significance of Newton's Metaphysics. *The Genesis of Hypothesis* is the realization of that plan. Burtt notes, "It will be impossible in this volume to develop an adequate doctrine of the categories in detail. *That must wait for a genetic analysis of the methods of science and philosophy, which is the third topic of the projected series.*" The word "volume" typed into the manuscript is scratched out and replaced with the word "essay", handwritten. Also scratched out is the sentence highlighted by italics. By the time Burtt was rereading and making corrections in the manuscript, he had scaled back his project.

The explanations of the categories time and causality parallel the one for space, but the treatment is cursory. "I pause here to refer briefly to the categories of time and causality and only to one fundamental matter in each case," he wrote. (GH, 111) The gist of the subsequent argument is that "time supplies an objective standard by which to judge succession" and "its ultimate basis is the flow of attention in its rhythmic pulses of arrest and transition." Time, like space, "is a habit of distinguishing between sequences as thus regulated and the unstandardized influence on our rhythms of attention due to impatience, precipitation (sic?), and delighted absorption." The category of time keeps us on task in pursuit of our purpose.

Causality is "the generalized need of dependable prediction of subsequent in terms of preceding changes...[causality thus] arises and supplies a unified principle of order among events. As unified, it is, like space and time, the product of generalized need; [and like the other categories emerging in our consciousness] under the pressure of need, it is, like them, a phase of the external structure of nature." His use of the word "phase" suggests that Burtt considered the categories to be some part of physical reality. Human experience is not entirely disconnected with the world as it is. What he wants to analyze is how we make sense of physical reality in this culture, through categories of mind, space, time, causality, quantity, etc. As if to clarify, he says, the categories are adopted to produce the "patterns of connections of experience imposed upon us from the outside, while being also, in their unity and totality, the product of mind." Even so, "space, time, and causality never construe with entire adequacy all the features of experience to which they appear to be relevant. There is always the margin of odds and ends left outside the realm they rationalize. Our experience of totality, especially as exhibited in the mystic's vision of the world *sub species aeternitatis*, does not quite harmonize with the demands of time as an all-pervasive category." And "there are various types of bizarre phenomena now and again knocking at the door of science which we hesitate to welcome because of our fear of the uncertainty in our basic categories which their entrance might bring, such as telepathy, phantoms of distant events, and prophecies of the future" (GH, 113-114).

The remainder of the manuscript offers some sweeping historical justification for the conclusions Burtt defended. From primitive man to scientific man passing through the intermediate stage of religious man, the same features of existence must be placed in an orderly framework.

What looks like evolution is not positivist. None of the ages of Western history, nor the frameworks of Eastern religion are to be assessed hierarchically. None is superior, none inferior. And further, Burtt notes, "To pass from one to another of these sets of concepts cannot be called either progress or retrogression." The fundamental structure of the world isexpressed in terms of dominant interests, and perhaps all are equally valid, objectively speaking. There is just one clue as to the progress of evolution in the whole of Burtt's grand scheme of explanation. It is that over time there emerges an ever increasing power to hold in the mind, simultaneously and effectively, a larger and larger range of experience. Whenever needs are generalized this progress is realized. As conscious unity pervades a vaster mass of detail a capacity for impartiality will arise along with the capacity to use critical analytical skills and separate consciously what we think of as scientific knowledge from the confusion and chaos that the world in itself really is.

Although *The Genesis of Hypothesis* is an unfinished work, the manuscript sheds light on important details in the evolution of Burtt's thinking after *The Metaphysical Foundations*. It addresses many of the questions he had been raising ever since his work on Newton. The abandoned manuscript is the watershed in Burtt's thinking. Important changes had taken place both in his career and in his family life in the late 1930s and early 1940s. Perhaps these factors prevented him from giving the work his full attention, or perhaps his interest simply changed. On Burt's own rereading, the projected "volume" became in his mind, a lessor work, an essay.

But he did keep the original manuscript through the years; he did not burn it in the bonfire on the banks of the creek near his house when he burned his other papers along with his deceased wife's confidential patient files. And, he came back to its central ideas in other unpublished manuscripts written in the 1970s and 1980s. One of these, the *Metaphysics of Empirical Theology*, is worth comparing because the conclusions reached in *Genesis of Hypothesis* through painstaking "proofs" are taken as axiomatic in the later work. There are fifteen interdependent points raised and itemized by number. The first is most interesting. Burtt leaves no doubt at all that his early idealism and wish to believe in the absolute has been weeded out in favor of his doctrines, derived from his philosophy of science, now applied to theology.

THE METAPHYSICS OF EMPRICAL THEOLOGY

1. Existence in its infinite fullness is unknowable and has no value or significance of any kind except as the focus of attention, perception and theory select from it those data which can be ordered into a system permitting far ranging inference and evaluation. In this way a universe is created, a home, so to speak, for the valuing consciousness of man indefinitely expanding, by creative communication from person to person, culture to culture, and age to age throughout human history.

The infinite fullness of existence has the value of being the inexhaustible source of new data; but the data have to be lifted out of it and ordered according to some theory created by the creativity operating in human minds before we have any knowledge or any value. In that sense existence has no value in itself apart from the creativity operating in living organisms and above all in human existence.

Theories about the infinite fullness of existence are metaphysical and can yield new knowledge by selecting data to fit new metaphysical theory. But this new knowledge is misunderstood when it is thought to be knowledge of the infinite fullness itself....

According to metaphysics here defined, there is no eternal structure of being which the mind discovers. What is discovered are the data which conform to the implications of a theory newly constructed by human imagination.

There is no Hegelian absolute giving structure to all existence and possibility, waiting to be embodied in human minds and the social order by the developments of history. Neither are there any eternal objects, as Whitehead would say, neither is there a limited number of categories, as Kant would say, prescribing, the changeless forms within which the mind must do all its thinking. Neither are the data of sense given to the human mind independently of theory and the selectivity of a ruling interest, as the English empiricists would say.

These interests selecting and ordering the data reaching consciousness may be practical or theoretical, aesthetic or mystical, scientific or metaphysical. But in all cases they are selected and ordered by operations of mind to meet the demand of communication, practical action, aesthetic drive, and the desire to know.

The mind is not free to select and order data as it wishes. The data are truly in existence and must be found where and when and as they truly are. But the fullness of existence is always more than these data selected out of it. This we know because new theories, new interests, and diverse ways of ordering and focusing attention can discover data previously inaccessible to conscious awareness. Different ages, cultures and persons live in different worlds. All this indicates that existence is an inexhaustible fullness out of which many different worlds can be created. But none of these worlds can

> be identified with existence as it is prior to the selections and orders achieved by theories directing focus of attention to find data to fit the theories. [7]

The rest of the essay is a pessimistic prophecy considering present-day trends toward self-indulgence and unwillingness to suffer. The saving anodyne is suffering in the same vein as Spinoza recommended in closing *The Ethics*--"all things excellent are as difficult as they are rare." "Experience of all great values requires sensitivity and sensitivity exposes to suffering."[8] In what could be taken as precautionary warning about trendy American-style Buddhism, Burtt wrote in point 11, "For [all the reasons above] any morality or religion or ruling commitment claiming that the goal of human life is to escape suffering is treason. It is treason because it blocks the way of man's salvation."[9]

Between 1921 and 1935 Burtt had undertaken a Humean program of disciplined skepticism and unrelenting empiricism to try to figure out how it is that mind interacts with physical nature to produce experience and how experience is filtered and reconstructed to formulate philosophical ideas, especially scientific and religious ideas. How the scientific reality of Western culture plays into the wider world picture became a well-worn path for Burtt to fine tune old ideas or to put forward new ones. Burtt traveled it many times after 1935 as if to enter the contemporary conversation in context only to take it somewhere entirely different. Sometimes the context was analytical and linguistic, if that was the preferred context of his intended audience; sometimes it was religious, idealist or cross-cultural. Burtt adopted a style demonstrating understanding of other view points or at least attempting to understand, before he presented his own ideas. He began to work his thoughts in and around the ideas of those he most wanted to convince.

After 1935 Burtt's seems to have changed his emphasis away from philosophy and history of science. In *Value and Existence* (1947),[10] Burtt deconstructed causality as nothing more than an invention of the Western mind, a manipulation of the experience of time for the purpose of predicting and controlling nature and events. But this was not the primary point of the article. His historical investigation of the concept of causality was simply an illustration that value considerations are at work when a generally accepted definition of a concept shifts. Causality served as an exhibit of how scientific and philosophical interests can shape our experience of a category. "Hume's conception of causality has seemed reasonable to many modern thinkers because it reflects an interest characteristic of a particular epoch of Western culture--the interest in accurate anticipation of future events."

Burtt's purpose in this article was to set his readers up to receive his own new philosophical method, "Cooperative Evaluation." This was a new brainchild of Burtt's and here he advocated it to decide "what definition of causality is the best one to adopt, in view of whatever competing definitions have come within our horizon and can be systematically compared with it....How is philosophy to go about the work of impartially comparing categorical definitions?"[11]

Over the entire decade of the 1940s, Burtt produced very little published work, except for the revision of his book on logic and a handful of articles advancing the practical value of his new philosophical method. As always he came back to points first raised in *The Metaphysical Foundations*. A network of presuppositions predetermines what thoughts can be generated. Individual "metaphysics" lie at the base of the unconscious thinker's logical structure. Such "metaphysics" arise from the cultural landscape, from personal preference or from the method used to solve the problem. During the 1940s Burtt thought less about the metaphysics of the natural sciences and more about the possibility of a scientific metaphysics for philosophy itself and for religion. The driving idea was to establish a scientific basis for categories of religious experience such as God, soul, salvation and so forth.

At Cornell, Burtt's major teaching assignment was a schedule filled with large undergraduate classes in the history of world religions. By then he had given up on the religious humanists of the early 1930s. Having drifted away from religion toward atheism the "Religious Humanists" had become the "secular humanists." Burtt's attention was turned almost completely toward the history and philosophy of religion now. He anticipated a synthesis between the religious philosophies of East and West, identifying and amplifying the best of both as a precursor to a world community. With this turn toward religious topics almost exclusively and his reluctance to get into the analytical and linguistic debates of the day, Burtt's talk of metaphysics was summarily dismissed by his Cornell colleagues as traditional thinking, out of fashion. This was not just unfortunate, it was inaccurate. Burtt had transformed metaphysics as the science which identifies and defines categories of human experience and logically explores the presuppositions upon which the categories stand or to which they lead.

To his own satisfaction, Burtt had identified categories within the scientific world-view. He accepted the world of experience as given by the culture, but understood it as a transitory mode of perception. Now, he was ready to go on with the rest of his own program, to wrestle with categories outside of physical experience, a vast, nebulous area occupied by philosophers and theologians of diverse stripes and colors. He wanted to examine the genesis of such categories as God, the soul, and dharma, to see how they define experience. He wanted to search for patterns in world religions and ferret out areas of agreement. Why have people constructed the spiritual categories? What are their most reasonable definitions? Can the history of world religions or the study of comparative religion illustrate any truths about spiritual experience? This is an elusive series of questions, for sure. As he hammered away at it a little at a time, he would come back over and over to safer, surer ground, where common understanding, at least in his view, was already established. This was the scientific method and the scientific world-view.

He always came back to one necessary presupposition and one only, that the universe is ordered and intelligible. Burtt made another assumption, too, in order to make his program work. He assumed that the scientific method is an essentially democratic open-minded affair, adopted by consensus among working scientists who exchange ideas freely and come to agree on the probable truth of tested theories. He simply took it as axiomatic that scientific truth is a socially agreed upon kind of truth. For Burtt the deconstruction of scientific terms was accomplished through the genetic method of history, which he had adopted from Columbia teachers, James Harvey Robinson, Charles Beard and John Dewey. The categories are reminiscent of Dewey's reconstruction in philosophy. They are the subjective factors needed to live in the world of experience given the perceptual apparatus we have and the time and culture of our births.

In certain ways Burtt's thinking anticipates Thomas Kuhn's. Kuhn and Burtt share:

1. An historical approach to the philosophy of science.
2. An emphasis on the metaphysics inherent within the scientific method.
3. The presupposition that scientific truth is more social consensus than actual apprehension of the facts of nature.
4. An appreciation of the role learning and culture have to play in the formation of all our concepts.
5. The denial that science progresses steadily toward the knowledge of anything that is true about nature.

Kuhn said that science gives aid in solving the ever-more complex puzzles the mind of man is able to propose. Burtt had said that science gives us a model for a method of cooperation in learning about the mind of man.

Burtt's work certainly did not attract the attention of the philosophers of science as Kuhn's work has done. This is because Burtt's intentions were perceived as different from Kuhn's and because Kuhn enjoyed the benefit of a new context for his ideas, the post-modern age. Burtt was seen as a religious thinker and dismissed as "soft" because he did not confront any existing controversy. Burtt made controversy the villain and cast about endlessly for points of agreement and synthesis.

Ironically, while taking as axiomatic his own unique spin on what the scientific method can tell us about nature, he seems to have been oblivious to the fact he was assuming something that has always been controversial. Historians of science have understood their discipline as writing the history of something whose philosophical and epistemological nature was known or at least was the business of someone else to define. Science refused to be historicized or post-modernized until Kuhn made the case for it in 1962 with *The Structure of Scientific Revolutions*. Perhaps, he, too, would have received scant attention had his volume not been published as part of the *International Encyclopedia of Unified Science*, a bastion of logical positivism and the "Whig interpretation" of the history of science. The logical positivists who read Kuhn there like a Trojan horse were the last enclave of whiggery in history writing and the strongest proponents that the nature of science and its scientific methods mirror the logic of nature itself. Burtt's point of view was never seriously entertained by the dominating group of historians of science, until Kuhn came along.

The power of attraction in Burtt's early work, beyond the extraordinary clarity of his writing, is the nonchalance and equanimity with which he undertakes the exposition of heretical ideas. No one else could challenge the canon with quiet calm quite like E.A. Burtt. The academic world loves less its peacemakers than its warriors who venerate the scholarship with their battles. The controversy over Kuhn's work brought Burtt ideas into the arena of debate in the same way controversy and Thomas Huxley's love of it brought Charles Darwin's ideas to the forefront although there had been a good many evolutionists prior to Darwin.

David Hollinger has pointed out that what Kuhn's *Structure of Scientific Revolutions* has done to the scientists cherished belief in a fixed, permanent natural order that can function as a standard for truth is just what Darwin's *Origin of Species* did to the doctrine of the design argument, toppling man from his special place in the universe. Hollinger says the antipathy to Kuhn comes from "critics [who] charge...that Kuhn is naive about how great and abiding is our need for cosmic anchors, for ideals of perfection unsullied by social, psychological, and historical functionality." He says that Kuhn's theory of science has "dispense[d] with the idea of a fixed, permanent natural order that can function both as a standard of truth in the case of particular theories and as a goal for the progress of science."[12]

Long before Kuhn, Burtt dismissed universal order as anything more than a necessary presupposition of the exact sciences. He was able to do this because he was already convinced that the age of the design argument had passed away and been replaced by a new world-view no more stable or permanent than the previous one had been. Most important, the new world-view was no more or less true than the previous one. Truth was out, skepticism was in, and Burtt was the new David Hume. It was no more heretical of Burtt to puncture the illusion of universal order than it was for Hume to puncture the design argument. Burtt showed how one had simply been substituted for the other without any better fit with "reality" or our chances for apprehending it.

Those who have marshaled forces against Kuhn have been the guardians of an independent universal order in nature and theory independence of method in science to reveal it. They have been those with faith in science to understand and control nature by mathematically cracking the code of natural order. Those who opposed Darwinian natural selection believed the complexities of organism design too supremely adapted to have arisen haphazardly. They had to have been designed and intelligently ordered according to Divine logic and Divine purpose. These problems never troubled Burtt, who simply accepted illogic as one of the features of the cosmos, or at least as one of the human experiences of the cosmos which has surely not been nullified by any human ideas or methods developed so far.

Genesis of Hypothesis is a bold effort to restore rationality after sweeping it all away by casting human perception and cognition together with the evolution of anatomical structures and locating the present human condition on an evolving scale. In this aspect the work is quintessential Burtt, an embryonic capsule of the later Burtt, whose early thinking steered a rambunctious path through all that was established, cutting it down by questioning and shaking up presuppositions. Rambunctious but not reckless, Burtt had faith that his clear-cutting methods would yield a new harvest and this is what can be seen in the philosophy of the middle and later years when Burtt became "transformed" by psychoanalysis in the late 1940s and early 1950s.

There is a mystical quality in Burtt's own brand of rationality which touches the Beautiful. Like the saints, he gave up what he treasured most, belief in human rationality, belief in the rationality of the world, and belief in God. His radical philosophy, Humean inspired skepticism, was really a sacrifice of faith shored up with the optimism that the human journey is not over. He gave up his belief in God in order to find the reality of God; he gave up belief in reason, logic and truth in order to find them again in their realities. At the end of his life Burtt found the meaning he had craved to find in philosophy in the spiritual life of man. The later Burtt is an inspiration, but it is the earlier Burtt, radical, rambunctious and penetrating that appeals most to the scholar.

Burtt's vociferous critics have centered their attention on his historical examination of early modern science, his characterizations of Copernicus and Galileo, or his research on Newton's mathematical formulae and methods. His bold attack on Newton's authority as any foundation upon which to frame a philosophy of mind or to pin a world-view has not received the attention it deserves. Burtt's motives, to challenge the reigning philosophy of knowledge, have been overlooked in favor of examining his interpretation of the facts of history. His place in the development of postmodern scientific understanding, as a new philosophy of mind, should be recognized, if only in the interest of making the history of the philosophy of science more complete.

CHAPTER 9

BURTT'S NEWTON AND THE DEBATE OVER THE RATIONALITY OF EARLY MODERN SCIENCE

When Burtt read Newton's works, available in print,[1] in the early 1920s, including Sir David Brewster's *Memoirs of the Life, Writings and Discoveries Of Sir Isaac Newton* (1885), he had no way of knowing that the conception of Newton he derived from his research would come to teeter precariously on the edge of a body of controversial scholarship several decades later. In the decades following Burtt's penetrating study, Newton, along with Galileo, came to represent high stakes for professional historians of science who saw themselves as the first line of defense in a battle to preserve and defend the mathematical realism of early modern science. The most prominent members of the newly evolving profession imagined the Scientific Revolution to be a complete break with the superstitions, religious beliefs, and magic of the past. For them the era of modern scince was enlightened by true and certain knowledge delivered by the work of science heroes. Central to the success of the heroes had been the mathematical method used to frame and confirm hypotheses by experiment. The key figures of the method were Galileo and Newton, with Copernicus and Kepler playing notable roles.

When Burtt exposed the "metaphysics" of the Scientific Revolution as that which stands between physical phenomena and mind's understanding of them he undermined mathematical realism. He wanted to relocate the principle of order, once reposited in God, now moved by modern science into nature and mathematical descriptions of nature, into man and define it as the human ability to become aware of the presuppositions of thought, and later to see "metaphysics" as the study of certain categories through which we make sense of experienced phenomena.

It can not be overstated that Burtt thoroughly accepted the primacy of science--as it is--without God. He wanted to reconcile science and moral philosophy in the manner of Pragmatic naturalism. What was original with Burtt was that he did not think that true knowledge would be possible without a completely new philosophy of mind. He did not think that science gives us certain knowledge, and he was, therefore, less concerned with our understanding objective physical reality than he was in searching for the universal principles of human understanding. He did not think that the fields of science, philosophy of science, and the history of science, as they developed through the middle of the twentieth century, held any promise in this direction. Therefore, he did not really have much interest in modern science as such. In searching for the "universals of human history" as indicative of the universals of human thought, he accepted science knowledge, admired science methods, and attempted to make his own methods scientific. He struck out alone on a path which led him to explore the history of world religions and world philosophies. He seemed to be looking for the least common denominator in human moral and social behavior and at last came around to think that such could only be revealed by identifying what men could agree upon rationally.

The question Burtt put to the history of modern classical science was this: Why is the main current of modern thought what it is? How did the modern problem of knowledge come to be? Just how did it come about that men began to think about the universe in terms of atoms of matter in space and time instead of the scholastic categories? His answer was NEWTON, or rather the uncritical acceptance of Newton's work with its "precise mathematical use for concepts like force, mass, inertia"; [and Newton's] meanings for "old terms like space, time, and motion, which had hitherto been unimportant but are now fundamental categories of men's thinking."[2]

Newton could never have achieved his revolution had he not drawn on the legacy left by Copernicus, Kepler, Galileo, and Descartes, who made the universe intelligible by making it mathematical. Newton built upon that legacy and made mathematical intelligibility universal in the so-called Newtonian synthesis. Burtt's insights are summarized in a few pages of *The Metaphysical Foundations,* titled simply *Summary of the Pre-Newtonian Development*. Key to his thesis on Newton are certain contributions to the metaphysical background Newton absorbed.

> Copernicus dared to attribute to the earth a diurnal motion on its axis and an annual motion around the sun, because of the greater mathematical simplicity of the astronomical system thus attained, a venture whose metaphysical implications he could accept because of the widespread revival in his day of the Platonic-Pythagorean conception of the universe, and which was suggested to his mind by the preceding developments in the science of mathematics.
>
> Kepler, moved by the beauty and harmony of this orderly system of the universe and by the satisfaction it accorded his adolescent deification of the sun, devoted himself to the search for additional geometrical harmonies among the exact data complied by Tycho Brache, conceiving the

harmonious relations thus laid bare as the cause of the visible phenomena and likewise as the ultimately real and primary characteristics of things.

Galileo was led by the thought of the motion of the earth and its mathematical treatment in astronomy to see if the motions of small parts of its crust might not be mathematically reducible, an attempt whose successful issue crowned him as the founder of a new science and led him in his effort to see the fuller bearing of what he had accomplished to further metaphysical inferences.

In Descartes the early conviction that mathematics is the key to unlock the secrets of nature was powerfully strengthened by a mystic experience and directed by his invention, that of analytical geometry. Could not the whole of nature be reduced to an exclusively geometrical system? On this hypothesis Descartes constructed the first modern mechanical cosmology. After Galileo:

The scholastic substances and causes, in terms of which the fact of motion and its *ultimate why* had been accounted for teleologically, were swept away in favor of the notion that bodies are composed of indestructible atoms, equipped with none but mathematical qualities, and move in an infinite homogeneous space and time in terms of which the actual process of motion could be formulated mathematically. Intoxicated by his success and supported by the onrushing Pythagorean tide, Galileo conceived the whole physical universe as a world of extension, figure, motion and weight; all other qualities which we suppose to exist in *rerum natura* really have no place there but are due to the confusion and deceitfulness of our senses. **The real world is mathematical**...[3] [and ready for Newton.]

These sweeping generalities recall Deweyan-type history with a broad brush. As more information has come to light on Copernicus, Kepler, Newton and especially Galileo, Burtt's history has been corrected, while his insights and conclusions have been revised. What is remarkable is the degree of accuracy Burtt achieved in finding his Newton and the influence his thinking has had on later historians of science through the decades.

Burtt saw classical modern science with its emphasis on quantifiable properties of matter and mathematical formulae as just one phase of the kaleidoscope of world views parading through history. He wanted to claim reality, insofar as we can know it, for the transitory world-views because these views are immediate human reality and essentially all we can experience. This did not diminish his making an accurate entry into the minds of historical figures he studied. Indeed, Burtt's genetic method demanded that close study be coupled with empathetic, intuitive inferences and clear historical insight to make his thesis sound. Burtt's sensitive, critical history of early modern science revealed the role of intuition, creativity and the irrational in scientific discovery. The result was that each of Burtt's scientific figures was arguably swept along toward an ever-increasing mathematization of nature by aesthetic or metaphysical preferences.

Although the contemporary fashion in the history of science still favors analytic philosophy and scientific reductionism as an index of intellectual rigor, a fair number of historians have broken away from an interest in Newton's physics and mathematics to explore more obscure questions about Newton, the man, and the historical context which raised his genius. Many of these questions were asked for the first time in *The Metaphysical Foundations*. What did Newton mean when he wrote that he "did not feign hypotheses?" How do Newton's rules of reasoning square with his science and his assertion about hypotheses? Did Newton succeed in keeping his science free of hypotheses, and if he did, how did he explain the phenomenon of gravity? Did Newton separate his mathematics from the phenomena they represented? Was he more of a Greek-style rationalist or more of a scholastic empiricist, a nominalist, who left the big picture to a Providential God? What were Newton's relationships with the Cambridge Platonists, especially Henry More and Isaac Barrow? What was the "sleeping argument" in defense of the existence of God? Why did Newton defend God? Was he simply trying to appease religious elements in his society or did deep theological conflicts lie at the center of his science?

Burtt's Newton was a conflicted seventeenth century figure, fully engaged in the theological and philosophical questions as well as the scientific questions of his day. With only one foot in the future, Burtt's Newton is a transition figure. Newton was only part modern scientist. He was part seventeenth-century theologian, too; part religious dissident, with links to the Cambridge Platonists and Arianism, and part renaissance natural philosopher, with links to Galileo. He was a strict empiricist and rationalist mathematician but also part magician, who believed in miracles and respected Providence. Burtt's Newton was a full-fledged voluntarist in spite of his scientific reputation.

It was Burtt's contention that Newton's enterprise was more than mathematical science, and that he separated the mathematics from the phenomena they described. This is what has rankled mid-century historians of science. Newton, according to Burtt, kept hypotheses out of science and this made it impossible for him to offer mathematical formulae for the nature of gravity, for absolute space and absolute time.

> Newton...was willing to grant the possibility of exceptions, but he was in no wise willing to grant that science was composed of hypotheses. Anything that is not immediately deduced from the phenomena is to be called an hypothesis and has no place in science, especially attempts to explain the nature of the forces and causes revealed in the phenomena of motion. Such explanations by their very nature are insusceptible of

> experimental verification. We know, for example, that certain motions take place in nature which we have been able to reduce to mathematical law, and regarding these motions as the effects of a certain kind of force, we call that force gravity. "But hitherto I have been unable to discover the cause of those properties of gravity from phenomena, and I frame no hypotheses." (Principles, II, 314) The ultimate nature of gravity is unknown; it is not necessary for science that it be known, for science seeks to understand how it acts, not what it is.[4]

On Burtt's interpretation, Newton's inability to offer exact mathematical formulae for the categories of space and time indicated logical inconsistency. Newton must have known as much, and he invoked God in order to resolve the philosophical rough spots. Just why Newton appealed to God is a matter for continued debate. That he did it, is beyond question.

Showing that God was indispensable to Newton's science gave Burtt the leverage he needed to pry positivism apart from its unrecognized presuppositions. At the same time he was prying mathematical realism off from Newton and this was his heresy among the twentieth century historians of science.

Burtt argued that Newton's science, including his descriptions of absolute time and space, were founded upon a presupposition of order in the universe, which is only the substitute for another presupposition--the existence and extension of God as universal order itself, God as creator of that order or maintenance man for that order. According to Burtt, Newton's description of motion depends on his description of absolute time and space but Newton, "did not make the distinction" between relative time and relative space, which are vulgarly regarded as sensible and different from absolute time and space. Relative time and space are related to relative motion, but have no meaning in absolute space. Absolute space would seem to correspond to the infinite universe where no reference point can be absolutely established. Without the reference points we have no logically defensible concept of absolute space; without the reference points, which are impossible to establish, we have no absolute order, only relative order, no absolute laws, only relative ones.

"Absolute space and time as thus understood, by their own nature negate the possibility that sensible bodies can move with reference to them--such bodies can only move in them, with reference to other bodies."[5] Absolute space and time are infinite and homogeneous. Therefore it is not possible to say if any body by itself is either moving or at rest in absolute space and time. Motion and rest are meaningless terms without a definite point of reference. This is the argument as Burtt wrote it.

The existence of absolute motion in this sense, that is, the fact that bodies change their distance-relations, and that in any direction and with any velocity, implies that there is *infinite* room in which they can move; and the exact measurability of that motion implies that *this room is a perfect geometrical system and a pure mathematical time--in other words, absolute motion implies absolute duration and absolute space.* Thus far Newton's mathematical method as applied in the Principia adopts and perfects the notions of space and time which had begun to undergo philosophical handling and by somewhat analogous consideration, in the work of More and Barrow. Did absolute space and time, as Newton proclaimed them, mean merely and precisely this, the conceptions would be logically unimpeachable and would deserve inclusion among the definitions and axioms which furnish the foundation of his mechanics, in spite of the fact that they are quite inaccessible experimentally. That motion is experimentally discoverable and measurable presupposes them. To this extent Newton has justification for these concepts, and the fact, so often observed by himself, that space and time "do not come under the observation of our senses", need not distress him, as an intelligent empiricist.

But thus far alone can we go with Newton; no farther. For note: absolute space and time as thus understood, by their own nature negate the possibility that sensible bodies can move *with reference* to them-such bodies can only move *in* them, *with reference to other bodies*. Why is this so? Simply because they are infinite and homogeneous entities; one part of them is quite indistinguishable from another equal part; any position in them is identical with any position; for wherever that part or position may be is surrounded by an infinite stretch of similar room in all directions. Taking any body or system of bodies by itself, therefore, it is impossible to say intelligibly that it is either moving or at rest in absolute space or absolute time; such a statement only becomes meaningful when another phrase is added--*with reference to such and such another body.* Things move *in* absolute space and time, but *with reference to* other things. A sensible centre of reference must always be definitely or tacitly implied.

Now it is clear that Newton did not feel this implication of the meaning of space and time or observe the distinction. For he speaks about the possibility of combining the motion of an object on a ship and that of the ship on the earth *with the motion of the earth in absolute space;* furthermore in many passages, both in the *Principia* and in the briefer *System of the World*, he discusses the question whether the centre of gravity of the solar system be at rest or in uniform motion in absolute space. Since in his day there was no way of getting a definite point of reference among the fixed stars, such a question is obviously unintelligible--the very nature of absolute space negates the possibility of its having any assignable significance. How, then, did Newton allow himself to fall into the error, and include such statements in the main body of his classic work.

The answer to this question is to be found in Newton's theology. To him, as to More and Barrow, space and time were not merely entities implied by the mathematico-experimental method and the phenomena it handles; they had an ultimately religious significance...they meant the omnipresence and continued existence from everlasting to everlasting Almighty God.[6]

The evidence came from Newton's *Principles* and from the *Opticks*. Strongest is the famous general Scholium concerning God in the second edition of the *Principia* (1713). "He endures forever, and is every where present; and by existing always and everywhere, He constitutes duration and space...He is omnipotent, not virtually only, but also substantially."[7] Using the *Opticks*, to which Newton had added the queries, taking up a number of unresolved questions generated by the *Principia*, Burtt argued that for Newton "absolute space is the divine sensorium. Everything that happens in it, being present to the divine knowledge--the divine consciousness furnishes the ultimate centre of reference for absolute motion."[8]

Burtt attributed to Newton the idea of God as "infinite knowledge and almighty will." He says that "as a kind of animism all real or absolute motion...is the resultant of an expenditure of the divine energy." God animates the universe in Newton's appeal to the Divine to keep the system from running down and to provide the conscious center from which all motion is referred.

Early in his thinking, Newton proposed a main "phlegmatic body of the ether"--a medium of transmission to account for mechanical changes. He also conceived of "various ethereal spirits" to explain propagation of motion, electricity, magnetism and gravity. Later, he revised this to a uniform medium with graduated degrees of density and grossness. According to Burtt, Newton did not deal with the ether in the *Principia* because he could find no accurate experimental laws to express its operation. And further, Burtt says, that Newton believed in the possibility of universal transmutations in nature--the spirituality of the cosmos was guaranteed by the original creation by God.[9] Newton ascribed ultimate causality to the ether and seemed to regard it as the direct operation of God, responsible for such phenomena as gravity and the production of bodily motions by the act of will.[10] By denying that gravity was inherent in matter Newton kept God in the material world.[11]

If none of this makes sense today, given our modern scientific context for Newton's laws, Burtt reassures us it is the scientific world-view of our age which has us confused. "Explanations in terms of pious reverence are not critically examined." The curious and disturbing thing for Burtt is that the whole illogical system (the scientific world-view) persists after being divested of its "theological excuse,"[12] that is after the Enlightenment thinkers dissolved God. The Newtonian concept depends on God. Without God, time can only be the immediate now. Duration can not exist when Newton's concepts are tried by strictly empirical tests. There is nothing to hold the points of attention together when the link between the knowing agency locked in the brain and the material world to be known can no longer be bridged by a spiritual continuity. This was the same point Burtt made in his 1926 paper on time as a category of perception, a paper that was an extension of his general argument in *The Metaphysical Foundations.*

In the Newtonian world-view the complete picture with its problems of knowledge rested upon the great authority of Newton and Newton's evasion of those problems by appeals to God. "All the links in the infinite scene were supplied by God."[13] Ever since Hume, Locke and Kant pried off the Deity the situation has presented distressing philosophical challenges. Bringing all this to light does not make Burtt a Christian apologist. The intention was to undermine the position of the positivists of his day by making the point that the philosophy, which later became the analytical philosophy insofar as it pretends to be based on Newton and Descartes, makes no logical sense without God. Therefore, the new realist or positivist philosophy was, at its base, a religious philosophy, unaware of its own theology. Newton and Descartes had included God as axiomatic. Hobbes, Locke and Kant had stripped away the theological underpinning without further examination of the presuppositions upon which their own philosophy was resting, if it depended on Newton. This does not mean that Burtt wanted to restore God to the analytic position. It means he was following Hume in being the consummate skeptic of the reigning philosophy of his day. Just as Hume used Newton to undermine the Design argument and subtract God, so Burtt wanted to use Newton to undermine the positivists and add God back into their equation. Burtt showed how Newton had kept God in the material world, not with the intention of putting God back into the world himself, but with the intention of exposing the whole construction of modern philosophy as inadequate in general.

To achieve this much and justify his interpretation of Newton's physics as metaphysical Burtt connected Newton to Henry More, the Cambridge Platonist, through Newton's teacher, Isaac Barrow (1630-1677). This was mostly an intuitive link on Burtt's part since he could not have known that Newton owned several of More's books, some of them presented by More himself with appropriate inscriptions. We know of only one meeting between More and Newton; it is described by More in a letter to John Sharp (1680).[14] Burtt, diligent researcher that he was, noticed parallels between Newton's thinking and More's argument in The *Immortality of the Soul.* It was a brilliant insight.

Burtt had directed his own "purposeful attention" to the subject at hand. He selected data, to accomplish two goals. The first of these was to describe a place and purpose for man in the cosmos. The second was to acknowledge change as the only truth we can know. By emphasizing Newton's voluntarism, but preserving his science, Burtt put Newton in the position of the scholastic nominalists such as Nicole Oresme, who concluded that many natural events, such as the fall of a stone, were completely incomprehensible to human reason. "Truly, rightly considered, these things are less known than some articles of faith. Therefore, this alone I know with certainty, that I do not know anything with certainty."[15] Burtt's Newton did not feign hypotheses, but referred everything to empirical observation, allowing that science does require the presupposition of an orderly universe until we have direct contrary experience. By building the universe on probability rather than certainty and by recentering order in the rational mind of man rather than in nature, Burtt reclaimed two important themes for man through his historical study of Isaac Newton.

He redeemed man in a scientific world by assigning him a new place in the universe as the intelligent, ordering principle, which presupposes pattern, constancy, and order, thereby making it so. Burtt would reclaim for man what had belonged to God through the design argument and, later, to the realists' mathematical formulae. Second, Burtt opened the way for new cosmologies to be built with greater freedom, by placing them in the conscious mind of man, referring them to probabilistic rather than certain truth, and by recognizing the inherent truthfulness of change.

Burtt had spelled out his intentions clearly at the beginning of *The Metaphysical Foundations.* By the end of his argument he had left no doubt as to the direction he expected his future philosophizing to take. It is the Newtonian categories as they developed in the immediate post-Newton era that he wants to reconsider in light of what they mean when mathematical realism is equated with universal order. By implication, this refers to the Divine presence. It has been misleading that some writers have incorrectly inferred Burtt's intentions to have been a desire to reinstate God in the universe or to rehabilitate of Christianity.[16] Because Burtt abandoned his epistemological theory of categories to turn his attention to Eastern religion some time during the 1940s the misperception has endured, but a truer reading of *The Metaphysical Foundations* can be had by paying attention to Burtt's own assessment of what he was doing.

> A penetrating study of post-Newtonian philosophers quickly reveals the fact that they were philosophizing quite definitely in the light of his achievements, and with his metaphysics especially in mind. At the time of his death Leibnitz was engaged in a heated debate on the nature of time and space with Newton's theological champion, Samuel Clarke. Berkeley's *Commonplace Book* and *Principles*, still more his lesser works such as *The Analyst*, *A Defense of Free Thinking in Mathematics*, and *De Motu*, show clearly enough whom he conceived to be his deadly foe. Hume's *Enquiry Concerning Human Understanding* and *Enquiry Concerning the Principles of Morals* contain frequent references to Newton. The French

> Encyclopaedists and materialists of the middle of the eighteenth century felt themselves one and all to be more consistent Newtonians than Newton himself. In his early years Kant was an eager student of Newton, and his first works aim mainly at a synthesis of continental philosophy and Newtonian science. Hegel wrote an extended and trenchant criticism of Newton. Of course, these men do not accept Newton as gospel truth--they all criticize some of his conceptions, especially force and space--but none of them subjects the whole system of categories which had come to its clearest expression in the great *Principia* to a critical analysis. It may be that their failure to construct a convincing and encouraging philosophy of man is due in large part to this untested remainder. It may be that many of the terms and assumptions in which their thinking proceeded were in their unanalyzed form essentially refractory to any such brilliant achievement.
>
> The only way to bring this issue to the bar of truth is to plunge into the philosophy of early modern science, locating its key assumptions as they appear, and following them out to their classic formulation in the metaphysical paragraphs of Sir Isaac Newton. The present is a brief historical study which aims to meet this need....At its close the reader will understand more clearly the nature of modern thinking and judge more accurately the validity of the contemporary scientific world-view.[17]

Clearly, Burtt was referring to a long-range plan he had in mind. Burtt's readers can miss his playfulness, his polished skepticism, and his intellectual exuberance in challenging the new realists and logical positivists of his era if they turn him into an apologist for a medieval God. Readers right up through the 1950s have wanted to see his iconoclastic arguments as thinly veiled religious apology, thereby stripping away their philosophical importance. The fact is Burtt was teasing the realists, so "vigorously committed to their own pet presuppositions," that they were blind to the illogic of the final consequences of their own conclusions.[18] His method was nothing but their own method unleashed back on their own shaky edifice. "Let us revert to the realist's procedure of determining the meaning of a concept by analyzing it into simple constituents fixed by the relation of presupposition,"[19] he once said. "Let us go realist all the way." The result was a remarkably new way of looking at concepts by stretching them to the limit. This is one of many reasons why *The Metaphysical Foundations* has an arresting effect. Burtt's extraordinary wit and love of the joke was turned into an integral part of a timeless lesson.

Considering the biographical material he had to work with, the many layers of deceit surrounding Newton's cherished image and the paucity of theological papers to study, how did Burtt find his Newton?

In 1924, Burtt believed that Einstein's science and the philosophy it would inspire would compete with the Newtonian view and replace it. The history of science as a professional discipline, with its own peculiar orthodoxy was still in the future, and Burtt certainly had no investment in keeping early modern science rational. From his vantage point, Einstein had already illustrated that the mathematical experimental methods lead to some curiously different concepts of space and time than the ones Newton had advanced. In 1924 Burtt was aware of no controversy in allowing Newton the full range of his genius and thus he acknowledged with equal validity all the religious treatises he saw listed in the Portsmouth collection and all which appear in Horsley's *Opera Omnia* right along side the scientific works. It was in Sir David Brewster's 1885 edition of *Memoirs of the Life, Writings and Discoveries of Sir Isaac Newton* that Burtt encountered a candid disclosure and discussion of Newton's religious interests and theological writings.

Brewster had come into possession of Newton's theological manuscripts and letters by way of Reverend Jeffrey Ekins, Rector of Sampford. Starting with the 1855 edition (the work was first published in 1831) there was a new preface in which Brewster declared that he felt compelled to discuss Newton's theology, and to bring it before the public, in spite of the fact that he and others who admired Newton would disagree with it on account of its anti-trinitarianism. In 1831, Brewster had written that Newton was a believer in the Trinity.[20] By 1836, he had found and examined the theological manuscripts. So carefully prepared were they that they had every appearance of having been intended for publication. Brewster listed all the theological manuscripts[21] and wrote, "It has surprised us that other authors should have regarded the study of the scriptures as incompatible with scientific research."[22]

Brewster had treated Newton's theology on its own terms, not relegating the theological work to Newton's later and, presumably, less productive years.[23] Brewster declared, "Newton believed the Scriptures were given to enable prophecy and prove that the world is governed by Providence."[24]

> [Newton] had been a searcher of the Scriptures from his youth, and he found it no abrupt transition to pass from the study of the material universe to an investigation of the profoundest truths...
>
> And this I know, that he was much more solicitous in his inquiries into religion than into natural philosophy, and that the reason of his showing the errors of Cartes' philosophy was because he thought it was made on purpose to be the foundation of infidelity. Thus Newton's research into scripture was to refute Descartes.[25]

Burtt certainly made a note of this. He was impressed further by Brewster's insinuation that Newton equated God with universal law,[26] that he understood universal order as a presupposition, and that universal order was not directly apparent in experience. He noted that Brewster interpreted

Newton as having believed that no design is obvious in the universe, no general law seems to direct planetary motions, and no regular system can be recognized in the chaos.[27] Further still, Brewster attributed to Newton the view that the book of nature is absolutely sealed except to the greatest of minds and so too is the book of Revelation.[28] Brewster seemed to be saying that Newton thought that the premises of science must be taken on faith as Scripture is taken. He says that Newton believed that false systems of religion could be compared to false systems of the universe, "but the false system proves the existence of the one that is true."[29]

Two threads of controversy have been pulled from Newton's theology--one out of step with the seventeenth-century Church of England, the other out of step with eighteenth-century Enlightenment science and some prominent mid-twentieth century conceptions of Newton, the man, and his place in the history of rational science. It turns out that Newton did hold Arian views, for which he could have been barred from holding any public position and possibly, his university chair, in spite of the Toleration Act of 1688. The Toleration Act made an exception against toleration for those who wrote against the Trinity. Brewster says that Newton was "hearty for the Baptists, Eusebians or Arians,"[30] but defends his freedom to hold any religious beliefs whatever and refuses to make apologies for Newton's having heretical religious views or for having his religion tied up with his science.[31]

Burtt studied Newton's own writing and Brewster's perspective on it. He concluded that:

> Newton's place in the religious unsettlement of his era would be an interesting topic for studious application. He was accused by the ultra-orthodox of being an Arian, apparently on ample grounds. Among other heretical suggestions, he wrote a brief essay on *Two Notable Corruptions of Scripture*, in each case the effect of his thesis being to cast doubt on the traditional assumption that the doctrine of the Trinity was taught in the New Testament. A strongly Arian flavour pervades most of his theological efforts, from which we shall take a quotation or two for another purpose, namely to show that religion was something quite basic to him and in no sense a mere appendage to his science or an accidental addition to his metaphysics. [32]

More important to Burtt's thesis than Newton's Arianism is where Newton's theology intersected with his science. According to Burtt's reading Newton had no inclination whatever to push God out of the universe. On the contrary, he especially engineered his science theories to make a place for God as part of gravity in order to keep Him very much in the universe his science described.

> Newton believed that scientific fact involved theism, but he would have been a theist had his scientific powers remained forever dormant. Newton evidently cherished a kind of religious experience, nourished largely, of course, by tradition, that was in the main detachable from the theism postulated as a corollary to science. This fact has its relevant bearings on his clear and continued conviction that the world of science [mathematical descriptions of forces and matter in motion] is by no means the whole world. (There follows a list of Newton quotations and citations from Brewster and from Horsley's *Opera Omnia* to illustrate the point.) [33]

This is the thread leading from Newton's theology that takes him far afield of the mid-twentieth century orthodoxy in the history of early modern science. Could it be that science is something less than purely rational?

When Burtt posed this question in 1924 the idea that science might not be completely rational was a proper enough idea to entertain, but within a few years, it would to become quite heretical to suggest. Two years after *The Metaphysical Foundations* appeared, A.J. Snow's study, *Matter and Gravity in Newton's Physical Philosophy*, argued that Newton had two hypotheses for the cause of gravity. One was a material ether, but the other and more pervasive in his writings, was an immaterial cause. If Newton reduced gravity to a physical explanation, God would be ruled out of the universe and this he could never accept. Newton had made the cause of gravity an immaterial force, the "ethereal spirit," identified it with the direct activity of God and thereby kept God in the world according to Snow's interpretation. In short, Snow's Newton was in concert with Burtt's Newton. It is not surprising that Burtt reviewed Snow's book right after it came out.

By the 1950s, things had changed. The canon of orthodoxy on Newton had now been laid down and can be seen in the work of Richard Westfall. In 1958 Westfall dealt with Snow's thesis, and therefore, indirectly with Burtt's, in *Science and Religion in the Seventeenth Century.* Westfall rejected Snow and argued that the early modern scientists, whom he called the virtuosi, were in conflict over the incompatibility of their science and Christian religion, especially the doctrine of providence. Even if God were identified with gravity, His job would be simply to uphold the general order of nature. Westfall's conclusion is that Newton's General Scholium, added to the 1713 edition of the *Principia,* was just an attempt to reconcile the mechanical universe and the God of Providence and not any acknowledgement of God's active place in the world. Overall, it illustrated only that the virtuosi were anxious to retain an orthodox God and keep God compatible with new theories, not to create any special place for Him.[34]

Burtt was not so much interested in religion per se as he was interested in how we know what we know and how we integrate experience with what mind brings to it. His purpose in examining Newton's categories was to understand the context in which Newton developed them and this was a religious context. He wrote:

> Religion was a fundamental interest to Newton. It dealt with a realm for the most part different from the object of science; its method was quite disparate, for its conclusions, in the main, were insusceptible of proof or disproof by scientific standards....[But for Newton] God was not detached from the world that science seeks to know...Although religion and science are fundamentally different interpretations of the universe, each valid in its own way, yet for Newton in the last analysis, the realm of science was dependent on the God of religion....Thus in spite of their incommensurable character and his considerable success in banning religious prejudices from his positive scientific theorem, the fact that God's existence and control was never questioned by the man who wrote almost as many theological dissertations as scientific classics had its strong and significant reactions on positions which he would have called purely scientific.[35]

> The rapid elimination of God, however, from the categories, rendered irreversible the projection upon modern philosophy of the notable problem referred to in the introduction [i.e. the disenchantment with the world] and yet racking the brains of thinkers, whose essential relation to the Newtonian metaphysical scheme can hardly therefore be passed over refer to the problem of knowledge. As long as the existence of a God to whom the whole realm of matter was intimately present and known, succeeded in maintaining itself as an unquestioned conviction, the problem of how man's soul, shut within the dark room of a ventricle of the brain, could possibly gain trustworthy knowledge of external masses blindly wandering in time and space, naturally became no terrifying puzzle--a spiritual continuity connecting all links in the infinite scheme was supplied in God....It was by no means an accident that Hume and Kant, the first pair who really banished God from metaphysical philosophy, likewise destroyed by a skeptical critique the current overweening faith in the metaphysical competence of reason. They perceived that the Newtonian world without God must be a world in which the reach and certainty of knowledge is decidedly and closely limited, if indeed the very existence of knowledge at all is possible.

> None of these keen and critical minds, however--and this is the major instructive lesson for students of philosophy in the twentieth century--directed their critical guns on the work of the man who stood in the centre of the whole significant transformation.[36]

A survey of the literature right up through the late 1950s shows very little direct concern with Burtt's descriptions of Newton's science or his treatment of Newton's theology with the exception of Edward Strong's rebuttals (1936 through 1970).[37]

During the 1920s the young radicals of Columbia University were discussing the role that the history of science should have in educating a whole population to live in a modern scientific culture. Two articles by Harry Elmer Barnes, who moved from Columbia to the New School for Social Research, and one by John H. Randall, Jr., are representative voices from the Columbia tradition. Soon they would be drowned out by the analytic types of the next decade. In 1919 Barnes had written:

> The only type of historian who is likely to give serious and sympathetic attention to the history of science is the member of that renegade and outlaw, but ever increasing group which has transcended archaic and illogical conventions and dared to view history as a record of human achievement conceived in its broadest sense as the progressive establishment of human control over nature and the increasingly more perfect adaptation of nature to human use....This position can best be substantiated by adopting at the outset the more recent anthropological theories regarding the origins of science. Our concepts in this respect were for years perverted by the vicious generalization of Frazer [The Golden Bough] to the effect that all of primitive science was to be found wholly in the realm of the occult, or, more specifically, in magical activities. [38]

In this climate of ideas *The Metaphysical Foundations* had been generated. After giving credit to the work of Franz Boas, also at Columbia in the 1920s, James Shotwell and James Harvey Robinson, whose research and methods "finally demonstrated the generic relation of religion and magic and...revealed the origins of science as proceeding primarily from...everyday secular and commonplace activities,"[39] Barnes summed up with the importance of teaching the history of science to undergraduates and noted that the history of science is part the Columbia course of study in Contemporary Civilization, a class taught to freshman by James Harvey Robinson, and later, by John H. Randall, Jr.

A decade later Barnes, then at Smith College, was writing about the need for a new religion based on scientific foundations.[40] His opinions seemed to anticipate the Humanist movement of the 1930s and Burtt's treatise, *Religion In an Age of Science.* Randall, too, was adding his voice in *Effects of Science and Industry on Religion.*[41] These men were not apologists for religion; they did not consider science and religion to be in conflict.[42] Rather they considered both science and religion to be in need of intellectual evolution, and they saw themselves as shepherds of that evolution. They were naturalist philosophers who dreamed of guiding civilization thorough the age of science, who saw science as a human endeavor with human consequences, and who saw religion as a human need. They refused to entertain orthodox views of either science or religion. Their philosophical rivals were the émigré analytics, who ignored both God and man by stuffing spiritual dilemmas under the rug and out of mind long enough to eclipse the naturalists and establish their own orthodoxy.

Over the 1930s and 1940s, philosophical concerns which had been emphasized by the sociological and epistemologically oriented naturalist pragmatists shifted. For them the study of the history of science was part of their program for right living through understanding the needs of the past and the solutions the past had adopted in meeting them. Burtt's book became a point of departure for the new wave of more formally minded Columbia naturalists, Ernst Nagel and more particularly, E.W. Strong. Later on Burtt was simply ignored by the professional historians of science such as George Sarton and Alexandre Koyré, who drew heavily on Burtt for inspiration, but emphasized individual science heroes for their precocious modernity and installed modern science as a glorious final achievement for the species.[43]

The beginning of the new realists' claim on Newton as one of their own can be seen from two volumes commemorating the Newton bicentenary in 1928. One is a volume prepared by the Mathematical Association of America, in which a paper by Burtt, *The Significance of Newton's Metaphysics*, appears. The other is *Sir Isaac Newton, 1727-1927*, a series of papers prepared for the History of Science Society. Both volumes contain a range of topics, more technical and scientific than historical. The second volume has a greater variety of papers and, also, a greater emphasis on Newton as a modern person. Two exceptions are a paper on Newton's alchemy and G. S. Brett's paper, *Newton's Place in the History of Religious Thought*. The essays were first presented as talks, delivered at the American Museum of Natural History in New York City in November 1927. They were introduced by Professor David Smith, President of the History of Science Society. In making the introductory address, Smith "[took] pains to make Newton 'modern' in his study of 'astrophysics', finance and religion." In the introduction to the published volume he wrote: "Professor Brett has shown that [Newton] was really one of the most influential predecessors of modern religious thinkers," meaning of course, atheistic or agnostic.[44] Thus was Newton stripped of his God by the early History of Science Society canon.

The History of Science Society came into existence in 1924. Burtt's particular interest, placing the history of science in a cultural context as one facet of Robinson's "new history" quickly became a minority view. Historians interested in science broke into two camps. One camp had adopted the German positivist program of Mach and Ostwald as college students. They tended to see science and the history of science as vanguard disciplines. The history of science would further civilization by furthering the development of science itself. These historians were frequently natural scientists themselves who promoted a "eulogy of heroic biography"[45] and whose scientific knowledge, pride and passion steered them into obscurantism as a polished style. George Sarton emerged from this camp to become the unquestioned leader of the Society. The Sarton message was "domination of science over, not integration of science with" the humanities. [46]

The side for integration of science with the humanities was represented by a number of James Harvey Robinson's students. Like their teacher, these historians promoted the history of science as one essential theme in a rapidly developing scientific-democratic mass culture. Some of the representatives are Lynn Thorndike, (*A History of Magic and Experimental Science*), Martha Ornstein (*The Role of Scientific Societies in the Seventeenth Century*), and Preserved Smith (*A History of Modern Culture*). E.A. Burtt, who was influenced by Robinson through John H. Randall, also belongs on this list. Their work was eclipsed by Sarton's domination in the history of science field.

Sarton's preference was for "remote periods, esoteric subjects, critical bibliography, and heroic biography,"[47] all of which continue to dominate the history of science even today through Sarton's journal, *Isis*. Sarton's vision of the history of science as encyclopedic, positivistic, linear expositions by science professionals did not inspire much interest or win very many disciples among general historians. It was not until the 1950s that the history of science became firmly established at the university level, and not until the 1960s and 1970s that a real explosion of interest took place.

For those reading *The Metaphysical Foundations* in the middle of the century, when science and religion were seen as orthodoxies in conflict, Burtt was characterized as "an outraged Christian humanist"[48] interested in the "welfare of religion" at the expense of "the welfare of science".[49] Reading about Burtt's Newton in Burtt's own historical context illustrates how far from the mark such an assessment is. The formalist temper of the age up through the early 1960s has not made it easy to read Burtt accurately. Now that Feyerabend, Popper, Kuhn, and the Continental postmodern historians have been assimilated, Burtt's history and his intention to work out a new philosophy of mind, based on the categories of the scientific world-view, make more sense.

In the 1950s when the history of science as a discipline was more widely studied at the university level Burtt was busy with other things. But his study of Newton, uncovering the curious mix of science and religion which Newton's purely seventeenth-century genius had woven together, was mired in one of the great intellectual controversies of the century. Philosophers of science in the United States, with their emphasis on formal logic, helped to shape the opinions of the historians even though the two professions hardly acknowledged each other. Beyond the fact that they did not meet at conferences or collaborate on research projects, they also denied the importance of one another's contributions. All the same, they held parallel views concerning the rationality of science.

Since the early part of the twentieth century, philosophers of science have identified rationalism with mathematical realism. When the history of science society formed in 1924, rationalism in science became identified with mathematical realism for the historians also. The historians conflated empiricism and Baconian experiment with mathematical realism, which they traced back to Plato, the Pythagorean tradition, the Mertonian scholastic tradition, or something arising for the first time with Galileo.[50]

Burtt's Newton did not fit this developing canon. Under Burtt's scrutiny, Newton turned out to be more than the isolated, hard-headed rational scientist generations of English-speaking college students would study in typical accounts such as the one offered by A. Rupert Hall in *The Scientific Revolution, 1500-1800: The Formation of the Modern Scientific Attitude* (1954). Hall is just one of a small army of science historians who abhor the irrationality of metaphysics mixed up with modern science. Naturally enough, Hall ignored Burtt's Galileo and his Newton.

Edward W. Strong, who was advised by F.J.E. Woodbridge, at Columbia, as Burtt had been, was more direct. Strong built the majority of his published work out of revisions to Burtt's Galileo and, later, to Burtt's Newton. In contrast to Burtt, and more in keeping with Woodbridge's own point of view, Strong attached mathematical realism to early science methods and weeded out metaphysics.[51] Strong's work is still frequently cited in conjunction with Burtt's by those who want to temper Newton's "metaphysics" with mathematical realism. Although he fought Burtt's view for almost forty years, eventually Strong came around to appreciate the significance of Newton's theology.

A. Rupert Hall, like the other successful historians of science of his type, has had no patience for philosophy. This is an attitude Hall shared with the late George Sarton, who preferred not to confound the literal, hagiographic history of science with questions concerning why some kinds of knowledge are more reliable than others or to question the validity of science findings beyond what the scientists themselves questioned and corrected with their "infallible" methods. Hall's book, *The Scientific Revolution*, was published in 1954 and revised in 1983.

Like Strong, eventually Hall, too, had to come to terms with Burtt's arguments.[52]

Two poles in thinking about Newton emerged in 1930s and 1940s and have matured in present-day scholarship. Louis Trenchard More produced *Isaac Newton, A Biography* in 1934.[53] More had looked at the same Newton papers on theology as Brewster had studied, but he had not considered them to have any scientific value and passed them off lightly. According to More, "Newton's philosophy and religion were two separate things, and he does not seem to have concerned himself with the problem of reconciling them."[54] Consequently, More's Newton does not mar the idealized picture drawn by the Enlightenment thinkers and preferred by the mathematical-rationalist historians in the United States during the 1930s through the 1950s.

On the other hand, John Maynard Keynes, a distinguished British economist, who became a Newton scholar, had a different view. Keynes had acquired about a third of a batch of Newton's papers offered at public auction by Sotheby's in 1936. The previously discounted manuscripts and correspondence had become the property of Lord Lymington, passed on to him from John Conduitt, who had married Newton's niece. The collection had been offered to Cambridge University some time in the late nineteenth century, but had been refused as insignificant. Some of the papers were purchased by Mr. Gabriel Wells and taken to the United States, others ended up in Jerusalem and were examined by Frank Manuel for his biographical work, *Isaac Newton Historian* (1963), *A Portrait of Isaac Newton* (1968) and *The Religion of Isaac Newton* (1974).

Keynes was committed to the idea that Newton's papers belonged in England. He thought it a shame that they should be sold at auction to become dispersed around the world. Consequently, he bought as many as he could and bequeathed the lot to King's College when he died in 1946. (He had offered the papers to Cambridge University, but the school refused them a second time.) In 1942, Keynes' delivered a lecture on Newton to the Royal Society Club. It appeared subsequently as an essay, *Newton, the Man*, printed by the Royal Society in *Newton Tercentenary Celebrations*. The image of Newton Keynes drew from the papers was altogether different from the modern figure the conventional positivist historians had been advancing. So, of course, it was a highly controversial one. Keynes revised Newton scholarship when he christened Newton, "the last of the magicians."

> In the eighteenth century and since, Newton came to be thought of as the first and greatest of the modern age of scientists, a rationalist, one who taught us to think on the lines of cold and untinctured reason. I do not see him in this light. I do not think that any one who has pored over the contents of that box which he packed up when he left Cambridge in 1696 and which, though partly dispersed, have come down to us, can see him like that. Newton was not the first of the age of reason. He was the last of the magicians, the last of the Babylonians and Sumerians, the last great mind which looked out on the visible and intellectual world with the same eyes as those who began to build our intellectual inheritance rather less than 10,000 years ago. Isaac Newton, a posthumous child born with no father on

> Christmas Day, 1642, was the last wonder-child to whom the Magi could do sincere and appropriate homage.[55]

Even in the late 1960s when Frank Manuel produced his highly psychological work on Newton, it was considered a grave and serious departure in the scholarship. Dominant strains in the history and philosophy of science, especially in the United States, could not admit soft subjects of a psychological or religious nature. In this context, it is significant that Keynes was neither an historian of science, nor a philosopher when he encountered the Newton papers. He had the freedom to read the Newton papers outside of the formal constraints and standards for rigor which stood as an artificial barrier to research by the so-called professionals.

There are two interconnected strains in the tight-lipped, argument against Burtt's Newton which have been carried on. One concerns Newton's science methods, especially the extent to which Newton was a rational scientist, pursuing modern mathematical science methods and modern experimental methods and purposes. The other concerns the reliability and rationality of science itself. The attitude seems to be that if Newton, as a king-pin of the modern method were found to be flawed by attitudes and interests which were clearly outside of that tradition, the whole enterprise of science might crumble. H. Floris Cohen has put his finger on the core of the anxiety when he recounts comments made by Paolo Rossi and Rossi's commentator, Rupert Hall, at a conference held on the island of Capri in 1974. Burtt's Newton is not the subject per se; rather it is the Hermeticist tradition and the debate over it, instigated by the work of Frances Yates' and the work of Thomas Kuhn, published in the 1960s.[56] The flavor of the debate is apropos. From it we get a better understanding of the realists' reception of Burtt's history of science.

Cohen notes that "Hall put the stakes particularly high" by responding to Rossi's interest in the hermetic tradition as he did. Hall commented:

> Rossi raises the issue quite simply--and he is right to do so; if the history of science is concerned with rational discourse between men, then the study of alternative models of discourse is certainly of auxiliary interest only; if on the other hand it is *not*, perhaps because of some special link between the 'pseudo-sciences' and the deepest levels of the human psyche, for example, then not only has the history of science as understood for the last three hundred years been a colossal fraud, but so has science itself.[57]

The younger Rossi, who had considered the Hermeticist tradition in connection with Francis Bacon, cleared the air by taking a middle road and disassociating himself with the most "irrational," least formal, and therefore most unacceptable historians of science.

> A recognition of the troubled waters at the origin of modern science, an awareness that the birth of scientific learning is not quite as aseptic as the men of the Enlightenment and the positivists naively assumed...does not imply either a denial of the existence of scientific knowledge, or a surrender to primitivism and the cult of magic.[58]

The heretics for irrationality were identified as Kuhn, Feyerabend, Roszak, Frances Yates and Rattansi. Farthest out on the edge was Frances Yates with her book on Giordono Bruno. She treats the problem of mind created by Descartes' distinction between primary qualities. The primary qualities are "real" for the mathematical realist because they can be quantified. The secondary qualities are less real because they do not lend themselves to numbers. When everything real is numbers, known by measuring, the world is entirely externalized. Mathematical science neglects the inner world, the world of mind. Yates had cited *The Metaphysical Foundations* in her complaint about the primary and secondary qualities and in her general conclusion.[59] The remarks Cohen reports from the conference, taken in this context, help to set the stage for A. Rupert Hall's later review of Cohen's book, *The Scientific Revolutions,* as well as Hall's treatment of E.A. Burtt.

Cohen placed Burtt in the company of three others, Eduard Jan Dijksterhuis (1892-1965), *Val en worp, (Free Fall and Projectile Motion* (1924), and *The Mechanization of the World Picture* (1950), Anneliesse Maier (1905-1971), whose research, published between 1949 and 1977 on science and scholasticism undermined Duhem's thesis; and Alexandre Koyré (1892-1964), *Etudes Galiléennes* (1935/36, published 1939) and *Newtonian Studies* (1950). For Cohen these historians are "the great four" of the "great tradition" in the history of science. In Cohen's opinion, it is Dijksterhuis and Burtt who independently established the fundamental themes, which would become the great areas of scholarship in the history of science. On Burtt, Cohen says:

> *The Metaphysical Foundations of Modern Physical Science* had a somewhat odd fate. It did not come out of any mainstream; nor did it create one for itself or was it ever made to fit any other. In its location beyond philosophical or historical currents or fashions it just represents that priceless thing; the individual thought of an individual thinker. The book need not be rescued from oblivion, though, for it continues to be reprinted and read to the present day. Still, I do not think that the significance of the book is widely recognized....There is a sense in which lonely Burtt's account is still very relevant to the course of present-day discussions of the Scientific Revolution. And what is even more: It is from Burtt's book that Koyré's concept of the Scientific Revolution probably found its origin.[60]

A. R. Hall had ignored Burtt's book in all his own writing until 1983 when he finally engaged him if only to take him to task for falling short of the mathematical realism which he found essential to the scientific enterprise. In dealing with the "great four" Cohen compares them all in separate sections. In his review of Cohen's book for *Nature*, Hall ignores Cohen's treatment of Burtt entirely, but mentions the three others in Cohen's "Great Four" category in one sentence. There can be little doubt where Burtt's Newton stands in Hall's estimation or why. Hall never went out of his way to mask a prejudice, even when the prejudice is out of fashion. He wrote:

> *Cohen's dictum that "To the extent that the history of scientific thought was cultivated at all, it was almost exclusively the province of the philosopher" can be accepted only if 'scientific thought' is interpreted as 'speculative thought'.*

And later, he complains, showing his bias that mathematics equates with rigor, that in Cohen's book, "The hard edge of scholarship tends not be felt."[61] To A. Rupert Hall's mind "hard science" and "hard scholarship" means defending mathematical realism.

If Burtt has been "lonely," as Cohen characterizes him, it is partly because of the prejudices learned by students from their teachers such as A. Rupert Hall. Hall's *The Revolution in Science 1500-1800, The Formation of the Modern Scientific Attitude* (1954), revised as *The Scientific Revolution 1500-1750* (1983), is significant as an index to the mainstream science historiographical treatment of Burtt.

In 1954 Hall made no mention whatever of Burtt or *The Metaphysical Foundations.* In his 1983 revision he speaks about new detailed studies since his 1954 treatment, which he now wants to take into account. Curiously, in his introduction he makes a reference to R.G. Collingwood's *Idea of Nature*, rather than Burtt's *Metaphysical Foundations* when the latter is the primary source of the idea he wants to discuss. Significantly, Hall chooses to remember Burtt this time, if only to disagree with him--once on Galileo and idea of force and then on Newton's concept of his own methods. It weakens Hall's point to use Burtt to make it when he fails to understand Burtt's commentary on Newton's method--that he separated his mathematical physics from his theology and that for Newton the mathematics were not more real than the physical phenomena. These points were important to Burtt's argument that the logical positivists had adopted Newton's methods while ignoring his metaphysics, metaphysics Newton freely acknowledged. It was the habit of analytic philosophy, invoking Newton to defend its method that Burtt challenged. This key point was missed or ignored by Hall, who needs a pristine Newton on which to peg his defense of mathematical rationalism. A.

Rupert Hall has never been interested in debates about the philosophy of science.[62]

Unlike Hall's work, or Edward W. Strong's, selecting out the historical figures and biographical details leading up to a culmination in rational science methods and real knowledge derived from them, all in support of mathematical realism, Burtt's book is a philosophical argument using history to highlight and suggest modern philosophical errors by treating universal order as a presupposition only. If universal order is only presupposed, and not actual, then the mathematical universe and mathematical realism require new qualifications. While Hall has been content to leave off with Newton's achievements as the pinnacle of Western thought and the end of the dark era of superstition, Burtt's early work had dealt with the history leading up through Newton as the beginning of a new dark age of unconscious metaphysics.

What was it that Burtt had said about Newton, also about Henry More and Isaac Barrow, that caused Hall to give him the cold shoulder, and finally, to feel compelled to engage him, albeit, without fully acknowledging the details of Burtt's argument or his reasons for offering it? Why did George Sarton, the father of the history of science discount Burtt? Why did Alexandre Koyré, whose own inspiration was Burtt, rarely mention him?

The answer lies within the history of the development of the History of Science Society and the preferences of its leader, George Sarton. While it is true that Sarton paid lip-service to the idea that the history of science is the history of civilization,[63] his conviction that scientific knowledge is certain knowledge destined to drive the future got in the way of his being able to entertain wider points of view. Sarton once wrote that the history of the quest for certain knowledge is progressive. The best he could do to widen and balance his perspective was to posit the possibly that China or India, and therefore, possibly, not the West had made the initial break-throughs. To recognize this much about other cultures, according to Sarton, will make the Western scientist "more humane."[64]

Sarton hoped to make the history of science, a subject that is necessarily technical, accessible to the non-scientist.[65] Koyré's treatment of the history of science suited Sarton on several levels. For one, Koyré, who was taking cues from Burtt, amply treated the cultural context of early modern science in a masterful way. Second, as noted by H. Floris Cohen, Koyré "knew how to sell his message" beginning with the *Etudes Galiléennes* (1939). The two immigrants, Sarton from Belgium, and Koyré, from France, understood American opportunity when they saw it and were willing to work hard at their endeavors. Cohen says of Koyré:

> His basic techniques were condensation, translation, and repetition....Koyré never tired of putting at the head of his numerous articles, as well as of his later books, a most emphatic statement of his core view.[66]

Koyré came to the history of science with a background in religion, and like Burtt, Koyré saw the down side of the Scientific Revolution. He acknowledged certain spiritual losses related to the shift in world-view when the cosmos as a qualitative, integrated whole was replaced by a universe of mathematical order in geometrical space. Therefore, there must be a different sort of key to Koyré's inclusion within the history of science canon at mid-century which will also explain Burtt's exclusion.

That simple key is that Koyré accepted the rational mathematization of the universe. Nature is subject to mathematical formulae, is controlled by universal laws and this knowledge is the legacy of the Scientific Revolution. Koyré did not challenge mathematical realism; he was not interested in epistemology, but rather in culture and cultural change. While it is true that Koyré was a little nostalgic for naive man, who believed himself to be part of the world when the world was whole, Koyré is a reflection of modern man, mathematical realist, but a little wistful about it. Burtt, on the other hand, was not modern, nor was he medieval as some have suggested.[67] He was, if anything, postmodern, although no label suffices. Burtt's thinking was a rare synthesis of the history of modern science, Enlightenment philosophy, and ideas at home in postmodernism. He melded past and future together as if making an attempt to skip modernism all together and pass right over Cartesian problems of knowledge and Russellian disenchantment with the world to come into an age which he might call, the age of human understanding.

It is ironic that, what Burtt said about Newton's mathematical science was mostly in line with what Koyré, Sarton and even A. R. Hall had to say. The following is from *The Metaphysical Foundations:*

> *For Newton...science was composed of laws stating the mathematical behavior of nature solely--laws clearly deducible from phenomena and exactly verifiable in phenomena--everything further is to be swept out of science, which thus becomes a body of absolutely certain truth about the doings of the physical world.* By this intimate union of mathematical and experimental methods, Newton believed himself to have indissolubly allied the ideal exactitude of the one with the constant empirical reference of the other. *Science is the exact mathematical formulation of the processes of the natural world.* Speculation is at a discount, but motion has unconditionally surrendered to the conquering mind of man.[68] (Italics Burtt's)

What could be a better rendition of the canonical approach? In fact, Hall quoted this passage in *Revolution in Science* (1983), if only to make the charge that Burtt did not go far enough to defend Newton's science from his practice, for Newton, according to Hall, "certainly never meant to compile an essay on the philosophy of science."[69]

If legitimate reaction to Burtt's work through the 1930s, 1940s and 1950s has been slow in coming, recent scholarship has made up for it. Now the connections between Barrow and More, Barrow and Newton, and More and Newton are well-known and acknowledged, although specialized debate still surrounds the extent to which Newton's thinking mirrors More's or synthesizes it into uniquely Newtonian notions.[70]

Burtt was on solid ground on Newton's relationship to Barrow. Barrow was not only Newton's teacher, but also his "intimate friend" and Burtt thought that through Barrow's connections with More, Newton came to understand More's belief in the extension of God into space.[71] "Newton agreed with More on the extension of God...and he certainly believed in ethereal spirits."[72] Burtt had studied More extensively. He had read More's letters to Descartes, his *Immortality of the Soul*, *A Platonic Song of the Soul, Divine Dialogues*, *Enchiridion Metaphysicum* and the *Opera Omnia* in four volumes. He says that More:

> ...extends his reasoning from the conclusion of an incorporeal substance in human beings [the soul] to the assumption of a similar and greater incorporeal substance in nature as a whole, for he was convinced that the facts of science showed nature to be no more a simple machine than is a human being. The facts cited in his further connections are the facts which had become the subject-matter of the most eager scientific study of the time, such as the ultimate cause of motion, cohesion, magnetism, gravity and the like. (Burtt cites More's Enchiridion). More notes that, although the sufficient immediate causes of motion can be described in mechanical terms, the ultimate reason why the parts of the universe are in motion rather than in rest cannot be accounted for mechanically.[73]

The irony of history for Burtt, is that since Newton could not prove the spatiality of God and the soul, he did not defend it himself. Therefore, the general picture of the universe and of man's place in it which went forth under Newton's name was essentially not his. It was the one already worked out by Galileo and Descartes and "in its most ambiguous and least construable form."[74]

Why did Burtt take such pains to restore God to Newton's world-view? One answer might be simply to give a more accurate picture of Newton; it certainly was not to censure Newton for his incomplete mathematical method, nor to criticize mathematical methods in the sciences.

Mostly, it was to illustrate the inadequacies of Newton's scientific world view as a general philosophy of knowledge and to offer hope that what it had become--positivism, materialism and disenchantment with the world--was not permanent. Philosophy could be redeemed, but dogmatic religion could not.

> Really, the notion of the divine eye as constantly roaming the universe on the search for leaks to mend, or gears to replace in the mighty machinery would have been quite laughable, did not its pitifulness become earlier evident. For to stake the present existence of activity of God on imperfections in the cosmic engine was to court rapid disaster for theology....
>
> Science moved on, and under the guidance of the less pious but more fruitful hypothesis that it would be possible to extend the mechanical idea over an ever wider realm, Newton's successors accounted one by one for the irregularities that to his mind had appeared essential and increasing if the machine were left to itself.
>
> While God was thus being deprived of his duties by the further advancement of mechanical science, and men were beginning to wonder whether the self-perpetuating machine thus left stood were really in need of any supernatural beginning, Hume's crushing disposal of the ideas of power and causality along another tack were already disturbing the learned world with the suspicion that a First Cause was not as necessary to any idea of reason as it had appeared, and Kant was preparing the penetrating analysis which frankly purported to remove God from the realm of knowledge altogether.
>
> Space, time, and mass became regarded as permanent and indestructible constituents of the infinite world-order, while the notion of the ether continued to assume unpredictable shapes and remains in the scientific thought of today a relic of ancient animism...The only place left for God was in the bare irreducible fact of intelligible order in things, which as regards the cosmos as a whole could not be quite escaped by Hume...and as regards the realm of moral relations was all but hypostasized by that ruthless destroyer of age-long theistic proofs, Immanuel Kant.
>
> Newton's doctrine is a most interesting and historically important transitional stage between the miraculous providentialism of earlier religious philosophy and the later tendency to identify the Deity with the sheer fact of rational order and harmony.[75]

Newton, who believed in animism and providence, became transformed by later philosophers into a modern positivist; subsequently he was adopted and transformed by historians of science into a mathematical realist. As a philosopher, Burtt was no mathematical realist. His contention that Newton was no positivist explains why A. Rupert Hall, who is a mathematical realist, but no philosopher, had so little use for Burtt. But that has not stopped Hall from making use of Burtt's philosophy of history recently in the changed environment of the late twentieth century. In his 1983 revised *Revolution In Science* Hall seems to have taken a lesson from E.A. Burtt, after all.

> We must grip the essential contrast between the whole modern world-view and that of previous thought, and use that clearly conceived contrast as a guiding clue to pick out for criticism and evaluation, in the light of their historical developments, every one of our significant modern presuppositions.[76]

Even if Hall uses Burtt's philosophy of history to examine the mathematical physics of the seventeenth century, he reaches a different conclusion by leaving Newton out. Convinced of the mathematical realism of Galileo and Kepler, Hall maintains that those two scientists believed that the "human mind understands the truth of a geometrical proposition as completely as the divine mind can do."[77] Hall asks, "Can either space or matter be infinitely divided; indeed, can the concept of infinity, essential to mathematics, be comprehended by the human mind at all?"[78] Hall argues that the seventeenth-century philosophers considered these questions and realized that their mathematics did indeed describe the world. For A. R. Hall mathematical realism was born with Galileo, matured with Kepler and Newton, and defines the Scientific Revolution. In 1983, Hall did not consider Newton's theology or any seventeenth-century theology to be an essential dimension of that era.

Within a decade, he had to reconsider. B.J. Dobbs, who died in 1994, made an eloquent apology for the Whig historians like Hall during a lecture at the 1993 Annual Meeting of the History of Science Society. Speaking about unconscious assumptions, reminiscent of Burtt, she said:

> *To my mind the issue of the proper interpretation of our scientific heroes has been the most pressing problem of all, a problem that was at least in part generated by the concept of the Scientific Revolution. I think the problem arises somewhat in this fashion: we choose for praise the thinkers that seem to us to have contributed to modernity, but we unconsciously assume that their thought patterns were fundamentally just like ours. Then we look at them a little more closely and discover to our astonishment that our intellectual ancestors are not like us at all: they do not see the full implications of their own work; they refuse to believe things that are now so obviously true; they have metaphysical and religious commitments that they should have known were unnecessary for a study of nature; horror of horrors, they take seriously such misbegotten ideas as astrology, alchemy, magic, the music of the spheres, divine providence, and salvation history....*

With specific reference to Newton, Dobbs asked:

> How could Newton retain his belief in a sun-centered cosmos after he himself had described an infinite universe? One cannot imagine that, mathematician as he was, he did not know that no center was possible for an infinite space.[79]

By 1994 there was an overwhelming body of new scholarship directly related to Burtt's thesis, at least as it concerns the history of science. (Although, Burtt's major point was actually not the history of science, or he would have pursued it. Burtt's point, as has already been shown, was the necessity for a new philosophy of mind--the so-called all-embracing philosophy of man.) A.R. Hall finally came around to consider Burtt in 1983. By 1990 he was engaging him more-or-less head on, as if Burtt's work were just then brand new, a full sixty-five years after *The Metaphysical Foundations* was first published. In the 1990s Hall published two works, *Henry More, Magic Religion and Experiment* (1990) and *Isaac Newton, Adventurer in Thought* (1992) and here he finally comes to grips the influence of the Cambridge Platonist, Henry More, on Newton's concepts of force, space and time. True, Hall avoids as much direct reference to Burtt as he reasonably can, relying for the most part, instead, on Koyré's accounts. What is important to a history of Burtt's philosophy and *The Metaphysical Foundations* is that Hall finally had to acknowledge the views Burtt advanced in 1924, views which became heretical, views from which he had recoiled throughout his career. Could Burtt's philosophy of history, once quoted by Hall, play any role in releasing Hall and others like him from the chains of orthodoxy they had been unable or unwilling to break? Hall wrote:

> As everyone knows, Isaac Newton in his *Mathematical Principles of Natural Philosophy* drew important distinctions between absolute time and relative time, *absolute* space and relative space, and it is hardly less generally believed that Newton derived these distinctions from Henry More.[80]

Those responsible for "this belief," according to Hall are, A. Koyré and E.A. Burtt. Hall is unconvinced, but he has to admit that:

> Isaac Barrow, the first Lucasian Professor of Mathematics at Cambridge and later Master of Newton's College, Trinity, expounded in his lectures ideas similar to those of More and Newton--in fact, his expressions are closer to Newton's later formulation than are those of Henry More. If it is untrue that Barrow ever exercised a formal responsibility for Newton's education at their College, it seems certain that the relations of the youthful Newton with the Lucasian Professor were closer than those with More, though the latter

> was from the same county and the same school. For Barrow was a mathematician, and though when a young man Newton certainly had a considerable interest in the sort of metaphysics in which More excelled, it was not to be his major preoccupation. It might be argued -- and is indeed implicitly argued by Burtt--that besides influencing Newton directly More also influenced him indirectly through Barrow.[81]

Hall goes on to thinly deny that Newton was significantly influenced by More in spite of a remarkably long list of More's books found in Newton's library.

> I do not, in fact, believe that More's highly imaginative, sometimes flamboyant, style of writing would have been congenial to Newton, although a number of his books were to be found in Newton's library. One, *Tectractys antiastrologica* (1681), purports to have been More's gift to Newton. Another was one of More's pamphlets against the alchemist Thomas Vaughan. Four works are More's studies of the prophecies of Daniel and John, a subject of intense scholarly interest to Newton; one of these too was presented by the author. Newton also owned a copy of the *Philosophical Poems* of 1647, which he had bound with *An Antidote against Atheism* (1653), and *The Immortality of the Soul* (1659). Copies of the *Antidote* and of *A plain and continued exposition of the several prophesies or divine visions of the Prophet Daniel* (1681) bear annotations in Newton's hand.[82]

It is easier for Hall to defend Barrow's influence on Newton instead of More's because:

> Barrow, though a theologian like More, was given to neither mystical nor metaphysical notions.[83]

Barrow represents mathematical realism. Barrow is an acceptable influence; More is not. Burtt's ideas had matured into a groundswell of scholarship so formidable that even A. Rupert Hall felt obliged to reconsider him, if only to make a brave defense of the old-time historiography. Burtt's interpretation of Newton was pioneering. Without any dogma or orthodox tradition to uphold, Burtt's history of Newton opened the field.

The Metaphysical Foundations of Modern Physical Science is an enduring work in the history of science because it is a classic of timeless philosophy. It is as an application of pragmatic philosophy that the book is most accurately read. Burtt's analysis of Newton's metaphysics outlasts the fashions because it stands on the hardest philosophical ground of all, ground that endures regardless of the popular mode in problem solving--it stands on the questions which are ever relevant--what is man? What can he know? And what is his relationship to the world?

In the 1980s, Burtt wrote a paper for private use, *The Metaphysics of Empirical Theology*. From it the philosophical sense of *The Metaphysical Foundations* can be better understood.

> Existence in its infinite fullness is unknowable and has no value or significance of any kind except as the focus of attention, perception and theory select from it those data which can be ordered into a system permitting far ranging inference and evaluation.[84]

In retrospect Burtt's interpretation was prophetic. Leaving Thomas Kuhn aside, by the early 1960s many scientists themselves were questioning the reliability of science knowledge, and indirectly, mathematical realism.

One example of this is a paper written in 1962 by John K. Wood, professor of physics at Utah State University. Wood had read Burtt's *The Metaphysical Foundations* and his *Types of Religious Philosophy*. He makes the argument that the concept of causality in science is a "presupposition" based upon the experiences related to statistical probability of occurrence. In a manner recalling Burtt, Wood asks:

> Are there really repeated patterns of events in nature, or is this a position to which we are forced by the very nature of our reasoning ability? It is quite true that science must depend on the recurrence of natural phenomena; but is this so because all events in nature are repeatable, or is it because these are the only parts we are able to understand?[85]

To illustrate, Wood describes a theoretically possible, but statistically unlikely, event that all the atoms in a crystal of salt might all move upward at once as a quirk of their constant random motion, thus causing the crystal to defy gravity and levitate. Were this to happen, would anyone admit seeing it and risk being thought absurd? An actual World War II event demonstrates the complexities of human behavior when faced with "miracles," events which break the pattern of assumptions about the nature of the world we live in. A seaman saw a flash of green light in the Pacific at sunrise. The event was spectacular, but the seaman did not report it and dismissed it as just a mental "quirk."` Years later he saw an article in the *Scientific American*, describing the phenomenon and only then did the sailor begin to talk about it. Wood wrote, "If a person does not believe in the presupposition of the repeatability of nature, he will not be interested in science but will turn to other areas of thought where this assumption is not important." It is human assumptions, Wood concludes, that control the structure of the universe.

> *Assumptions are made for many different reasons and become so basic in contemporary thought that when...explanations are proposed that are*

> *counter to the ones we accept, doubts are raised about the explanations rather than questions concerning the assumptions on which the new explanations are based....*
> The presuppositions of both science and religion are...positions which have been taken to explain experience without any reasonable proof of their ultimate truth.[86]

In *Thematic Origins of Scientific Thought, Kepler to Einstein* (1988) Gerald Holton explains the value of mathematics to science.[87] Its adoption as an impartial criterion has rendered science statements uniformly understood and reduced metaphysical propositions, "masquerading as empirical or analytical ones." Mathematics has made science strong and successful and is a major reason why science has progressed so rapidly since the seventeenth century. Mathematics has controlled the "discourse of science." But mathematics is not the whole story.

> These successes do not, however, hide the puzzling fact that [mathematics] excludes an active and necessary component that is effective in scientific work...It neglects the existence of preconceptions that appear to be unavoidable for scientific thought, but are themselves not verifiable or falsifiable.[88]

Holton wants to examine "how the individual scientific mind arrives at the products that later can be fitted into the established theories or how science grows and changes." This is what Burtt did in his 1935 manuscript, *Genesis of Hypothesis*. Perhaps such matters are beyond the realm of the scientists or even the philosophers, but Holton thinks the only place to begin is "through studies concerned with the nature of perception, and particularly of the psychological development of concepts in early life."[89] And just as Burtt had done, Holton recommended the anthropological approach, looking for recurring "themata" in the preoccupation of scientists. In the 1930s, Burtt had attempted to devise a new metaphysics of science categories to explain how scientists arrive at hypotheses and confirm them.

Burtt had glimpsed the promise of these ideas in *The Metaphysical Foundations*. Many readers have misunderstood Burtt's purpose in targeting Newton's metaphysics and have described it as an attempt to bring God back in all His medieval glory. But Burtt's use of Bertrand Russell contrasted with Dante was just a convention of the era, a signal to certain of his readers that his thesis was intended to address the tensions of the modern age. As we have seen, his actual purpose was to challenge the status quo in philosophy, not to bring back the old time religion. But, without the benefit of the two early papers from 1926 and 1927 or the unpublished manuscript, *Genesis of Hypothesis*, Burtt's real meaning might be lost behind contemporary presuppositions concerning the "eternal" conflict between science and religion.

The fact that religion and science were not so much in conflict in the seventeenth century as they were mutually interdependent has been hard for thinkers after Burtt to comprehend. So much of modernity since the 1930s has been bound up with anti-religious and pro-scientific dogma that Burtt's insights, interweaving these two, have been hard for strictly modern readers to appreciate.

Burtt was surely enmeshed in the science/religion conflict but he had no thought to roll back history, nor did he want to. In fact, Burtt wanted to use the history of Newton's development as an empiricist and mathematician to lift the veil of illusion which he believed surrounded physical science and the positivist philosophy. By exposing its vulnerable presuppositions he implied its consequent limits. Once the contemporary thinking was unmasked and shaken, the time would be ripe for a new era of synthesis between perception and experience, and new theories of knowledge would emerge. Here is where Burtt's interests really lay. Burtt's examination of Newton's religious beliefs bound up with his scientific thinking in *The Metaphysical Foundations of Early Modern Science* was a first lesson in humility about the claims the positivists were making.

In contrast to the historians of the 1950s, historians of the 1980s and 1990s recognize that Burtt's statements about Newton's theology are modest. It has even been suggested that Newton's enthusiasm to make mathematical formulae which would offer accurate predictions of the behavior of matter had a Scriptural counterpart. Newton hoped to find the keys in the language of *Revelation* that would enable him to read the Biblical prophesies and make predictions in the same way he could read the natural world in the language of mathematics to make predictions.[90]

Newton's unorthodox manuscripts were not unknown. They were examined in the eighteenth and nineteenth centuries by his biographers and other authorities, who decided that they were unimportant to Newton's science achievements and might, if published, even detract from Newton's image as the hero and symbol of the age of science. In the 1940s studies by H. McLachlan brought Newton's theology to light and by the 1960s the scholarship was reflecting it.[91]

Present day interest in the controversy is specialized into several camps ranging from those defending the strict rational Newton as the only Newton worthy of study to those contending that Newton belonged as much to the tradition of magic in a world ruled by a Providential God as he did to the Enlightenment Age.[92]

CHAPTER 10

WHAT HAPPENED TO *THE METAPHYSICAL FOUNDATIONS* AND BURTT'S INTEREST IN THE HISTORY OF SCIENCE?

While Burtt's influence continues to be a strong current in modern historiography of science, the origin of this current has been obscured by vigorous intellectual fashions, such as the trend toward analytic philosophy, artificial standards for mathematical-type rigor in science historiography, and an investment in mathematical realism as a guarantee for scientific certainty. These trends became cemented in ivory towers and turned into an institution with the Society for Unified Science, established in the 1930s. H. Floris Cohen is right when he says that the significance of *The Metaphysical Foundations* is not generally recognized.[1] Although the book continues to be reprinted right up to the present time, it takes its readers somewhat by surprise, in that they, like Professor Cohen, can not place the book in any tradition. As we have seen, the cradle for Burtt's revolutionizing ideas was Columbia University in the 1920s, a place where James Harvey Robinson taught the new history, John Dewey taught naturalism in philosophy and all the Columbia radicals embraced guiding the new age as their social duty as philosophers. Part of what happened to that movement, besides the wave of analytic émigré philosophers, is explained by the twentieth century backlash against, what Morton White has described as "anti-intellectualism in America."[2]

According to White, an anti-intellectualist will "press the claims of the heart and the hand against those of the head, or he may think of intuition as a superior faculty to be distinguished sharply from that employed by the mathematician or the experimental scientist."[3] The anti-intellectualist is opposed by the "philosophical analyst of reason," whose faith in pure reason as the most reliable method of problem solving and surest means to truth, makes him suspect even of "empirical anti-intellectualists,"[4] such as the American Pragmatists, William James and John Dewey. In the war games of intellectual aggression and counter-aggression described by White, empirical anti-intellectualists achieved recognition for their claims that some truths can only be known by sense experience. They were successful entrenching their position firmly into the realms of the pure rationalists.

According to White: "They forced a reconstruction of the idea of reason itself" and this, he claims "represented a turning of the tide." The result was that reason became more broadly conceived to include at least one area beyond pure mathematics. This is the climate in which Burtt's *Metaphysical Foundations* was conceived and born and this is why Burtt described Newton as a strict empiricist, but not a mathematical realist. Unreason and disorder became a possibility in a world shorn of the presupposition of order by acknowledging that Newton's third rule of reasoning came under his fourth rule, that is that empirical observation might overturn universal law at any moment.

At this point truly "aggressive anti-intellectualism" emerged demanding even greater concessions to the notion that there exist many other modes of insight transcending mathematics. At first this special insight was advanced by rigorous argument and was seen as belonging only to the "chosen few," who had it by special gift. They were the priests, metaphysicians, theologians, and sages. But such selectivity does not mesh well with populist American ideology. Subsequent democratic reform expanded the group with special insight to include "natural man" in his various guises. Thus anti-intellectualism in a democratic society threatened to became no less than the supra-rationalism of the pre-scientific age. The door was opened to include the passions, emotions and feelings as valid methods for establishing truth and knowledge. Burtt had always had an interest in psychology and after his period of psychotherapy he attempted to integrate it with his philosophy. But, in his case it was an attempt to identify the psychological overlay and separate it from truth in the manner of Deweyan reflective thinking, rather than admitting Bergsonian elements.

This is the perceived danger in American philosophy as it flowed from William James to John Dewey and E.A. Burtt. It has been the business of American philosophers since 1930 to check and contain the anti-intellectualist tendencies in any established doctrines based on the assumptions "that there are no more than two ways to arriving at reliable knowledge of truth" and these are "logic and empirical observation."[5] While *The Metaphysical Foundations* in and of itself is on solid ground, it could easily become a point of departure for less formal thinking. White adds, "we may think of the history of philosophy as a story of misguided efforts to present other [misguided efforts]."[6] Or to prevent misguided efforts in misguided ways.

E. A. Burtt and the other young radicals in New York City in the 1920s were exploring new means to reliable knowledge and new definitions of truth to support their new methodologies. Part of what happened to Burtt's new history and philosophy of science, is that Alexandre Koyré and others purged it of its reformist agenda and made it acceptable within the formalist program. Koyré took the wind out of Burtt's sails and diverted it to his own free-flowing history of science. The sheer energy and determination with which George Sarton and Alexandre Koyré forged the new discipline left Burtt behind as he turned more toward philosophical inquiry than toward historical inquiry. The result is that *The Metaphysical Foundations* speaks to the critical historian of science and invites correction. It has hardly spoken at all to the philosophers of science for whom it was intended since they have steered clear of strict empiricism even as an extension of logical argument for fear of coming too close to the anti-intellectuals.

As Burtt began to emphasize the philosophical aspects of his Ph.D. dissertation, more and more, his reformist intellectual nature collided with his psychological make-up as concilliator. The analytic philosophers as specialized professionals rejected any duty to society as a whole and began to regard their lack of participation in the world as a virtue and measure of the serious importance of their work. Burtt's interest turned toward the philosophy of religion, an area with far more freedom for speculative thought than the history and philosophy of science and more potential to link theory and practice. With the serious study of religion on the wane in the United States, there was not a large community of philosophers with whom Burtt could commiserate.

An article by Stephen Toulmin for *Daedalus* in 1977 offers some insights concerning the differences between U.S. philosophy of science and the rest of the Western world which are especially useful in trying to come to terms with the historical context surrounding *The Metaphysical Foundations*.[7]

During the 1930s, 40s and 50s philosophy and history were separate disciplines, which guarded their self-imposed isolation from each other, especially in the United States. "The main tide of formalist philosophy of science, generated in Vienna in the years before Hitler, overleaped Britain and flowed directly to the United States."[8] There was no "orthodoxy" in the philosophy of science in Britain in 1950, but in the United States, the situation was different. Toulmin says that U. S. Philosophy of science influenced the historians of science by way of the "Unified Science" movement of the 1930s. Under the influence of the German émigré philosophers, Carl Hempel, Herbert Feigl, Rudolph Carnap, Hans Reichenbach, H. John von Neumann and Philip Frank, "all genuine problems in the philosophy of science were to be construed as problems about the 'logical structure' of the sciences, rather than about the psychology of scientific discovery or about the historical evolution of scientific concepts." This point of view was adopted by some formally minded pragmatists such as Ernst Nagel, who was a few years behind Burtt at Columbia, a student of Morris Cohen and protégé of F.J.E.

Woodbridge. Woodbridge, too, would have been sympathetic. Woodbridge advised Nagel and Strong as well as Burtt in their graduate studies. Nagel and Strong had no use for the historical process of discovery, such as Burtt presented in *The Metaphysical Foundations*. "The psychological vicissitudes of individual scientists, or the collective histories of scientific groups--were matters for the behavioral and social sciences, and had nothing to do with the philosophy of science."[9]

This attitude became so deeply entrenched that over the next generation there was a negative program of prohibition against confusing formal or logical issues, which were considered rational, with empirical matters, which were considered irrational. Toulmin remembers that "whether in history, psychology or the sociology of science, [the edict of prohibition] was widely observed even among people who were otherwise quite far from the logical empiricists in their thinking. To be convicted of the 'genetic fallacy'--as having muddled historical origins with logical justifications--was to risk being laughed out of court, and the same went for the parallel offenses of 'psychologism' or 'sociologism'.[10] Toulmin's purpose is to chronicle the changes which have occurred in the history and philosophy of science since the history of science "began to take off in the United States around 1950, [when] a general agreement quickly developed among the scholars directly involved about the proper topics of research within the field." It seems they focused on issues internal to the intellectual development of the various sciences. They were interested in the evidence and arguments which led to certain discoveries, the exact replication of experiments and who might legitimately claim primacy in science discoveries. They took care to steer clear of questions about the social context, economic motivations, and the like, deemed "external context" and "intellectually soft." Toulmin, who lived and worked through this period, writes in a frank and revealing sentence, "So, young graduate students whose imaginations began wandering further afield quickly became sensitized to those subtle academic pressures summed up in the lift an eyebrow and the interrogative murmur of the single word, 'unsound?'"[11]

George Sarton, founder of the History of Science Society, held a positivist's point of view and did not think that philosophy of science contributed to the advance of science knowledge. For him, according to Toulmin, "the scientist must approach the observed facts of Nature in a simple, unprejudiced frame of mind, and those facts would then 'speak for themselves.' State your question; do an experiment; get your answer--that was the straight-forward message, and the history of scientific thought was a record of good, rational empiricists seeking to follow this rule, in the face of interference from bad, irrational theologians and metaphysicians."

Philosophers of science and historians did not collaborate. They went parallel ways with a common commitment to an empiricist conception of scientific method. "If the philosopher's task was to establish the formal organon of science, the task for the historian was to produce 'rational reconstructions' of the scientific achievements of the past...This meant accepting the classic texts of natural science as intellectual 'masterpieces,' or 'holy objects,' and considering only the rational procedures and arguments by which their creators had been able to formulate and justify them." Since this was the orthodox methodology of the 1930s and 1940s[12] *The Metaphysical Foundations* must have been a controversial book for historians of science and their students. Nevertheless, it has been consistently read by general historians and philosophy students. It has been continuously in print since its first publication in 1925, and since 1932 in the revised edition.

Judging from the present day attention being paid to the historiographic issues Burtt raised back in 1924, it seems that *The Metaphysical Foundations* has endured through the decades as a subliminal tradition. Especially in the United States the book has represented a perennial counter-point to the establishment, all the while preparing minds for Thomas Kuhn's revolution in the historiography of science and for the decline of Sarton's view. Sarton had thought that philosophy has only a "corrupting" influence on history and on science, but today, "most philosophers of science would...agree that the whole story about the organon of science cannot be told in the formal language of inductive logic alone. Even those who remain committed to the program of developing a formal theory of scientific confirmation [about the reality of the natural world]...would acknowledge that its relevance and application to the actual natural sciences can be grasped only with the help of historical examples."[13]

The International History of Science Congress met at Ithaca, New York in 1967. Burtt had retired from Cornell by then and whether or not he attended is unknown. It was there that historians and philosophers of science were prompted to examine each others' work after Kuhn had confronted issues in the history of science in highly philosophical terms, reminiscent of Burtt's *Metaphysical Foundations of Modern Physical Science*.

Even today there are elements within the profession which treat Burtt as a heretic, Lorraine Daston, for example. But, on the whole, the tide has really turned among prominent Newton scholars; orthodoxy and dogmatic formalism are out. The tightly focused scholarship of the past has served an important purpose, preparing the ground for the wider panorama of today. Now there is a deep context into which the new information can be fitted and more speculative thought about Newton's work is better grounded. E.A. Burtt was simply ahead of his time in the 1920s and 1930s. Not finding a sympathetic audience within the professional societies, he turned away from the history and philosophy of science in the 1940s.

The professionalization of the history of science as a discipline was carried on in the United States by George Sarton. During the period between the wars, Sarton set out to install the history of science in the expanding American university system. His vision included university departments with special chairs devoted to it, a history of science society, and the journal, *Isis*. Koyré's account of the Scientific Revolution, based on Burtt's, appealed to Sarton. Koyré's energy and prolific scholarship along with Sarton's own encyclopedic accounts laid the groundwork. Taken together, Sarton and Koyré stood for a humanistic approach to science, rendering it in both philosophical and historical contexts, but the vision lost its steam, as the next generation of science historians took over. The Halls, A. Rupert and his wife, Marie Boas Hall, helped to skim the speculative philosophical context from the history of science diminishing its humanistic side. They preferred to give the facts in labored accounts, putting those facts in historical perspective, of course, but making the reporting as matter-of-fact as a science text. This style held no appeal for Burtt and it must have been one of the factors which drained his interest away.

Subsequent editions of Burtt's textbook on logic, *Right Thinking* (1928-1946), reflect an evolution toward a more positivistic approach to the history of science in Burtt himself.[14] In the 1946 edition, in a chapter titled, *How Science Progress*, Burtt had transformed his historical figures to comply more closely with the canonical profiles put forward by A. Koyré , George Sarton, E.W. Strong and especially by John Herman Randall. Failing to cite his own work in the recommended bibliography for the chapter, his detailed sketch of the history of early modern science seems to have been drawn from his old friend's textbook, Randall's *The Making of the Modern Mind*. This was an odd thing, since Randall had depended so heavily on Burtt's Newton in writing that very text.

In 1940, Randall had put forward a new thesis on the influence of Aristotelianism on Galileo at the school at Padua for which he became quite famous.[15] In this thesis, Randall denied any contribution of neo-Platonism to the formulation of Galileo's mathematical "book of nature." Burtt had claimed in *The Metaphysical Foundations* that Galileo's mathematical structure for the world had a religious basis as a fundamental aspect of his own thesis, and that the mathematical uniformity of nature is a religious and therefore metaphysical premise. For Randall, the school at Padua was Averroist, without theological basis. Randall portrayed Galileo as a culmination of a progression in self-criticism taking place in the Aristotelian science taught at Padua and thus, "the science of the Italian universities could progress steadily in self-criticism to the achievement of a Galileo."[16] Not one to take on academic quarrels and hair splitting defenses anymore, Burtt skirted around the disparity between Randall's opinion and his own. He wrote:

> Galileo inherited this self-critical Aristotelian tradition of Padua, but the faith that the universe is a uniform structure of mathematical law took especially firm root in his mind. His main chosen task was to demonstrate the validity of this faith, and since most of his contemporaries--the more orthodox followers of Aristotle--rejected it, he had to devise a general procedure for convincing them.[17]

At the end of the chapter there is the barest hint that a new revolution in science, the Einsteinian, has replaced the Newtonian world view. And with this Burtt officially, if weakly, backed off from his priority in *The Metaphysical Foundations*. As he told it later, he had other interests. As to the long-standing friendship with Randall, Burtt and Randall had already broken over another issue, Burtt's allegiance to psychoanalysis and related criticism of the Humanists. By 1967, when a volume of essays was collected and presented to Randall on his sixty-fifth birthday, Burtt was "out of the loop." The editor of the volume notes that E.A. Burtt contributed a paper to the project, but that it was rejected.[18]

A good many people asked Burtt in his later years why he gave up his priority as one of the great historians of science. Typical of him, he passed off references to his early success. Most graduate students around Cornell did not realize that he had written the book they were reading for Max Black and those who were aware could hardly believe that the quiet, demanding teacher they knew could have been the author. Burtt was too modest, too humble and too peaceful a man to want to enter into hard-nosed philosophical battles with anybody. His success as a thinker and satisfaction as a man came from extending his own understanding, rather than hair-splitting. In the last analysis, it was a personality conflict with the discipline that made him glad enough to leave it to people like A. Rupert Hall and L. Pierce Williams, history of science professor at Cornell during the 1960s. The analytical school in the philosophy department spilled amply into the history of science there.

The Metaphysical Foundations probably was not intended as a purely objective, presupposition-free historical exposé of the illusion of modern thought. It strikes me a glee-filled joke on the realists, their own method turned back on them. What was new to philosophy was Burtt's notion that metaphysics is an examination of the categories of human perception, conditioned by experience in the environment, human needs, and eventually by assumptions or presuppositions about the way the world is constructed. Burtt thought that a scientific analysis of the categories of modern science would reveal a structural pattern and a new philosophy of mind acceptable to both the behaviorists and the idealists. Burtt's underlying contention was that experience and perception are woven together interdependently by the thinking processes of mind. He imagined that an understanding of the categories of experience, modified by perception, needs, and presuppositions, would enable him and other philosophers to pursue their true vocation in the world, guiding human civilization toward an ever more complete experience of the world. Burtt did not imagine that the fullness of human existence was encapsulated in scientific understanding of physical reality. The fullness of human potential would be a spiritual experience. What was innovative about Burtt's religious thought was his pursuit of the idea that categories of human spiritual experience might be identified just as space, time, causality and the other physical categories were identifiable. He thought that the spiritual categories must be universal experiences and this led him to disqualify God as a category, because not all people have an experience of God.

It was the quest for an understanding of the spiritual categories that fueled his interest in the religious humanists. Since they were not receptive to including spiritual experiences in the Manifesto I, and later moved away from spiritual concerns altogether, pursuing morality and ethics only, Burtt lost interest in the movement.

Turning instead to psychology, not in an intellectual way at first, but for personal reasons, he became enchanted with it for awhile. It is unclear how much of Burtt's interest in psychology grew out of his intellectual concerns and how much was inspired by his second wife, Marjorie Murray. It certainly took on a spiritual significance for him. During the same period he was becoming more knowledgeable about world religions, especially Eastern religion, which blended with his experiences in psychoanalysis and his interests in the philosophy of mind.

At Cornell there was not much of a receptive environment for the direction Burtt's thought was taking, but he pursued it all the same, charting his own path with persistent humility, fortified determination, intermittent cheer and occasional enthusiasm. Burtt reassessed his own ideas as rigorously as he evaluated others. Finding them flawed sometime in the late 1930s, it was impossible for him to settle down to a persistent point of view beyond the need for human understanding, for compromise, and for advocating his own method of achieving that much. Personally, Burtt pursued the life of the spirit, finding the fullness of human existence in his personal relationships, his political activism, and his religious sympathies with the Quakers.

During the 1960s, the tide in philosophy and in the historiography of science turned, and Burtt began to think about his early work once again. He wrote a few papers for private use, making references to the historical method and the analysis of presuppositions exemplified in *The Metaphysical Foundations*. Still, he was an island in a churning sea of competing ideas that lacked the very things that might have provided useful keys in analysis and understanding.

1.) A willingness to examine unconsciously held beliefs, presuppositions.
2.) A real historical perspective, uncontaminated by or chained down by philosophical fashions.
3.) A cooperative spirit of inquiry and
4.) A love of truth and respect for the fallibility of man.

These were the true foundations of *The Metaphysical Foundations* and they became the foundations of Burtt's actions, of the way he lived his life.

Where did these ideals lead? They led, in time, from the life of the mind to the life of the spirit, to service to humanity, and to the examined life as the life well-lived.

The public intellectual in America has all but disappeared at the end of the twentieth century. While many educated and thoughtful people read and study a variety of popular psychological and quasi-philosophical works, they have no public leaders to guide them. In some ways the dreams of the Columbia radicals have been realized, since more people than ever have the benefits of higher education, but there is little in the way of real inquiry or public introspection. In this age of mass media, where every man is his own intellectual authority and guide, public intellectuals, like Burtt, are gone. It would be hard to imagine a common context like the one Burtt used in *The Metaphysical Foundations*, quotations from Russell and Dante on God, to open any general discussion from which the majority of the listeners would immediately draw inferences and conclusions making it possible to lead the discussion to a new place. Today's culture is a hodge-podge of ideas and opinions without standards for quality, or methods for agreement. More and more it is the press which tries to assume the role left open by the young radicals of the progressive era. The professional historians and philosophers have retreated to their ivory towers to carry on specialized discussions among themselves.

Burtt never took much heart in this kind of thing and this is why he turned away from the history of science and from philosophy as it was practiced. He made concessions, like the one to Randall's Galileo in his later writing, but he never gave up on the mission of the radicals to guide mankind through the scientific world-view with its disenchantment with the world on to something more spiritually and intellectually satisfying. It just became harder and harder to carry out the vision.

CHAPTER 11

BURTT'S PHILOSOPHY AFTER 1935

There is something ethereal about Burtt, his personal manner and his philosophizing, after the mid-1940s. That is because during the years of the Second World War and afterward Burtt joined the small brotherhood of spiritual seekers who have passed through the dark night of the soul. In 1943, Burtt began to work with Dr. Alva F. Gwin, a Freudian psychiatrist in Albany. The period of psycho-therapy was long and the effect was deep. Not only did Burtt's private life change, his whole manner of philosophizing was transformed. Some of the process showed up in his publications from the period.

In the Autumn issue of *The Humanist*, 1945 and again in the Winter issue of the same year, articles by Burtt appeared. Burtt's articles were followed in the Spring issue by a response from John Herman Randall, Jr. The collection of articles, including Randall's response, reflects not only Burtt's spiritual crisis but also the changed emphasis his philosophy would take from then on. Taking the tone of a preacher, not a philosopher, Burtt's articles are critical of religious Humanism for ignoring the benefits of psychology. The first, *Does Humanism Understand Man?* attacks the essential optimism of the movement in the face of the specter of mankind's "sick soul" in need of redemption. By this time Burtt had been in psychoanalysis for about three years and it seems he wanted to restore "sin" to the humanist lexicon, defining it through the "scientific" theories of psychoanalysis as "unconscious" motives and drives that can lead to evil. The remarkable thing is the conviction with which Burtt delivers his argument. He claims that salvation comes through psychoanalysis and only those who travel the psychoanalytic route have the fitness to philosophize and lead. World War II is taken as evidence that "psychoanalytic theories of human nature... show, definitely and incontrovertibly, both how deep-rooted in feeling and controlling in action our blind emotions and antisocial desires are."[1] For example, Burtt wanted to argue that the psychological phenomenon of "projection" had been at work in the unforgivable treatment of naturalized Japanese citizens.

Burtt complained that no religious humanist had stepped forward to avert the West from "the horror and world-wide catastrophe." In *Humanism and the Doctrine of Sin* he picked up the theme again. Burtt's tone in this article recalls his preaching at St. Paul's Church in 1920. He is as a man "born again." Born again, not to religion and the reality of God, but born again to the reality of self-awareness and the tendency to self-deceit, born "in the fire of penetrating self-understanding." Through psychoanalysis, Burtt had seen the darkest side of his own human nature. "Were we able to look at these dark forces honestly, we should see that they control a far larger part of our personality than we have ever dreamed," he wrote. "To expunge them is a long, arduous, agonizing process in which the feeling of contrition plays at most a very partial role."[2]

John Herman Randall wrote a scathing response to Burtt in the Spring of 1946. In some ways, Randall's reaction was an apology for Burtt as his long-time friend. But, overall, Randall is incredulous at where Burtt has wandered, and attacks him for his "humanized, evangelical Methodism--or Arminianism." For himself Randall chooses instead "Niebuhr's naturalistic Calvinism." Where Burtt is pleading salvation for every man through psychoanalysis, Randall is no way convinced that psychology is the answer. Instead "those who deeply feel man's impotence and helplessness, in themselves and in others, will inevitably turn...to those forms of religion which can promise supernatural grace; and this, Mr. Burtt['s psychoanalysis] is not prepared to offer." No, Randall says, "Humanists should remain true to their own distinctive genius; they should continue to appeal to those who see no way out of mankind's difficulties except through social intelligence, education and a resolute will--whatever will be the ultimate inadequacies of these human resources."

At last, Randall gives up on the possibility of arguing sense into Burtt. He says, "It is bootless to try to shake the faith of those to whom has been given such confidence." Referring to Burtt as "he who has been twice-born" Randall says that Burtt "knows whereof he speaks. But the hardened reprobate may wonder whether Mr. Burtt is not still caught in the toils of 'that quite unwarranted rationalistic optimism about human nature' which has the 'blithe confidence' that knowledge--honest self-knowledge--can of itself bring us salvation. Those whose sense of human inadequacy may not be quite so desperate, and who have therefore a little more perspective and patience, may well have a keener realization of the limitation of mere knowledge--even self-knowledge--to make us whole." Sarcastically, he adds, "Unfortunately God has hardened the hearts of the reprobate--including, I suspect, those of most humanists. They seem never able to attain such a state of grace. They must seek salvation elsewhere, and by other means. Let us hope that the elect may have pity on us, and not exclude us utterly from such joys." Leaving no doubt that by "the elect" he means Burtt, Randall adds, "As history has abundantly shown, such a faith can profoundly move men, as it has moved Mr. Burtt. But can it move men--many men--to the point where they are fit to assume the obligations of 'complete liberty'? I fear that Mr. Burtt, even as [St.] Paul, will find many Ephesians amongst the redeemed." [3] This is in reference to Burtt's assertion that after all are saved through psychoanalysis there will be widespread self-understanding and self-discipline, at which time man is free to "love and do what [he] will."

Why had Burtt turned against the Humanists? Back in 1933, when Burtt was first contacted by the Humanist Fellowship group, it had appeared to be a party of religious liberals, seeking a sounder basis for man's spiritual needs than that provided by traditional religion. This was a problem dear to Burtt whose philosophical and theological interests had been centered on the idea that traditional religion was inadequate in the age of science.

Many pragmatists in the 1920s and 1930s were hopeful that humankind's ability through intelligence and the power of reason would lift people out of "despondency and doubt of God's loving Providence by confidence in the power of human intelligence and co-operative good will to become its own Providence."[4] *The Humanist Manifesto* was intended to bring a public challenge in the name of naturalism to the supernaturalists, whose beliefs were founded on revelation rather than reason and science. The challenge was intended as the hopeful response to the fundamentalist Christian revival of the era, which it set out to stem. For the Humanists traditional religion had become a disappointment in an age of science. It was simply a matter of time before thinking people everywhere abandoned faith altogether and they, like Burtt, foresaw that that would only add to man's "disenchantment with the world." The idea, then, was to develop a new religion, but scientific and progressive with modern ethical values. The Humanists put their faith, not in the power of God, but in the power of the intelligence of man.

They signified their interest at the time of the first *Humanist Manifesto* by calling their movement "religious humanism" and for Burtt this was the significant thing. As time went on and various proponents took up the banner it was called naturalistic humanism, scientific humanism, and ethical humanism, depending upon the particular emphasis of the individual. The group always stressed inclusion and was repelled by religious dogma, therefore opening itself willingly to many points of view and many interpretations. At first, this was another strong attraction for Burtt, but after a while the ideology became dogmatic and stale to him.

In time the humanists moved away from religion altogether, and thus, they grew away from Burtt, who stayed the course through the period. Indeed, there is not much difference between the argument Burtt made in *Religion In An Age of Science* (1929), which attracted the Religious Humanists to him in 1933 and a paper he wrote for *The Journal of Liberal*

Religion in 1944, *How Are Theological Conclusions Demonstrable?* The later argument has a profoundly Christian tone, but is none-the-less logically identical to the 1929 argument. Compare this from the conclusion of *How Are Theological Conclusions Demonstrable?* to his position in *Religion In an Age of Science*, where he had argued against the reality of God on the basis of the fact that God is not demonstrable in universal experience.

> *Theological propositions are only demonstrable, in the fashion of scientific conclusions, to those whose experience of the world has already become such that they can interpret it in full harmony with the theological categories and the doctrines expressed in their terms....To them the demonstration is superfluous...To others, neither the reality of God nor any other theological doctrine can be demonstrated by argument...And it may easily appear insulting as well as impotent....*
>
> Proof of theology can never be an affair of verbal argument alone, nor even of such argument clothed in vividly appealing symbolism. It...requires exemplification in conduct...[5]

The case is substantially unchanged. Theological conclusions must be universally demonstrable; they must be part of experience. The only scientific demonstration of religion is through right action.

> The only pertinent argument here is action; here is an obvious field in which the truth reigns: Those who talk do not understand and those who understand do not talk. So make my world, as far as in you lies, more like the kind of world I would expect it to be if God were real. Increase my evidence by your conduct...[6]

Corliss Lamont became an eloquent writer and spokesperson on behalf of the Humanist philosophy after the group dropped its religious associations. Lamont was not an original member of the Chicago group, although his signature was solicited by them, he did not respond to the letter. Lamont graduated from Harvard in 1924, studied at Oxford and eventually came to Columbia University where he studied under Woodbridge and Dewey, just as Burtt had done five or six years previously. Naturalistic Humanism he wrote, "rejects all forms of supernaturalism, pantheism and metaphysical idealism, and considers man's supreme ethical aim as working for the welfare of all humanity in this one and only life, using the methods of reason, science and democracy for the solution of problems.[7]

Lamont says he came to Humanism only after seeing that the *Humanist Manifesto* was signed by "such eminent philosophers as John Dewey and John Herman Randall, Jr." This strengthened [his] conviction that mature individuals, relying on reason and science, should adopt Humanism as their way of life."[8]

But after World War II, American thinkers were more sober about professed ideologies than they had been after the period of expansion in the 1920s. For example, Point fourteen of the original Manifesto[9] had called the attention of McCarthyites and anti-communists, prompting one original signer, Curtis W. Reese, to write:

> The controversy about point fourteen appears to me to be much ado about nothing. By no stretch of the imagination can point fourteen be made to support Soviet Communism. We must not allow our anti-communistic attitude to sway us out of accord with the world-wide trend toward a more socialized economy.[10]

World War II had changed the American perspective. Public intellectuals like John Dewey were growing older and their voices did not hold the authority they once did. By 1953, both Burtt and J. H. Randall were backing away from the Humanists, but for different reasons.[11] Burtt thought he had a better program of his own and Randall thought the Humanists had squandered their opportunities.

In the March/April, 1953 *The Humanist*[12] published a symposium issue which included the original *Manifesto* with *An Historical Note*, written by Raymond Bragg, one of the original group members and a key organizer of the Manifesto project. There was also an article by Roy Wood Sellars, who had written the original draft of the Manifesto back in 1933. Sellars' article was titled *Naturalistic Humanism: A Framework for Belief and Values*. Also appearing in the issue was a collection of statements, which were "responses of the surviving signatories (and some selected others) to the questions: 'How has the manifesto stood the test of time?' and 'If a new statement were to be prepared today, what changes should be made?'[13]
Burtt wrote to the symposium committee as follows:

> The word "Humanism" still comes closer than any other to representing my general philosophic position. But I've come a long way since 1933. And to square my present thinking with the details of the Manifesto would require that I complete, right now, what I hope to work out in a book some ten years in the future. Many issues of basic importance are involved.

Quite likely Burtt is referring to his book, *In Search of Philosophic Understanding* (1965), which emphasizes psychological theory as the foundation for understanding why men engage in philosophical argument. By 1953 he had already published his steps for philosophic understanding which emphasized psychological methods in three places--*The Philosophical Review*

(1946); *The Journal of Religion* (1947); and in a new chapter added to his revised *Principles and Problems of Right Thinking, Right Thinking, a Study of Its Principles and Methods* (1946).[14]

By the time of the symposium, Randall, too, had turned away from Humanism. He responded to the request for the symposium in more detail than Burtt had done. He was concerned that Humanism had failed to satisfy its early promise.

Second, Randall wanted Humanism to offer meaning to people in the form of religious symbols of some kind, more than just mere words and dreams. He complained of:

> *...insufficient recognition of the need of imagination in religion, and of the role of religious symbols....Humanism should face seriously the very difficult problem of creating more adequate imaginative symbols. It should at least recognize the need even if it cannot yet satisfy it.*

The symposium had grown out of controversy among various points of view, competing heavily within the schools of philosophy adopted by the various original signers. As viewpoints had matured, the first Humanists had diversified and there "was rather vigorous complaint that the editorial policy of The Humanist was too pragmatic in its orientation." This was the position of the philosopher Arthur E. Murphy who chose to drop out.[15]

Other controversies involved Roy Wood Sellars, who was a critical realist and therefore, in epistemology, anti-Dewey. There was also a rather sharp conflict among those who were more positivist in outlook, Charles Morris, John Dewey, and Arthur Bentley. There was also conflict between Bertrand Russell and John Dewey.[16]

While this could imply a lively debate around Humanist philosophy, it might also be interpreted as a reaction against the prominence Humanist ideas were gaining worldwide. In 1952 the International Humanist and Ethical Union had incorporated, establishing international headquarters in Utrecht, Netherlands and the "Declaration of the Founding Fathers" had claimed Ethical humanism "is...a faith that answers the challenge of our times."[17]

It is true that John Dewey and John Randall, Jr. did sign the manifesto as did E. A. Burtt, however, it should not be concluded that any of these men were ever active members of the original group who drafted the document, except for the suggestions made by Burtt. In fact, Dewey had signed without making and comment, and has therefore been unjustifiably linked with *The Manifesto* in exaggerated contexts. Edwin Wilson wanted to correct this misperception in his history of the Manifesto and brought it up under the subheading, *Misuse of Manifesto I*

> Dewey was not a principal shaper of Humanist Manifesto I but simply one of its signers. In fact, Dewey made no editorial comments at all. His name is constantly invoked only because he was, beyond question, the most famous signer...[18]

The period of Burtt's psychotherapy coincided coincidentally with his loose affiliation with the Humanists. Personal events (marital separation), coupled with world events (World War II) and the protracted period of psychoanalysis acted together to undermine Burtt's faith in reason and logic. Psychoanalysis taught him something profound, poignant, and immediate. It brought Burtt face to face with the darkest side of human nature.

His interests changed after that, some thought dramatically. Psychology, Eastern philosophy and religion engaged his interests and his academic energy flowed only for them. He lost old friends. The rift with John Herman Randall, Jr. was very bad and Burtt entered a period of loneliness. Analytical philosophy had been fully installed at Cornell, and that separated him from his university colleagues. Overall, Burtt's recognition had slipped since the years right after *The Metaphysical Foundations* and the years of the Chicago School of Pragmatism.

Matters continued to worsen at home, too, until he separated from his wife, although they did not divorce. Burtt's life was in turmoil for a decade. Then, in 1951 the clouds began to lift. He divorced Mildred Camp and married for a second time to Marjorie Murray, a child psychologist, who had herself undergone a Freudian analysis.

By 1951, E. A. Burtt was becoming a different sort of man. This is something reported again and again by those who knew him through 1930s, 1940s and 1950s.[19] During the 1940s his confidence in himself had lagged along with his confidence in the faculty of reason altogether. At the same time a persistent faith in the power of psychoanalysis to transform souls had swept over him and turned him into an evangelical Freudian. Infusing a distinctly religious meaning into psychoanalysis, Burtt compared his own transformation to the discipline of Confucianism and Buddhism.[20] He left the history of science and the philosophy of science behind. He abandoned his interest in a theory of categories of perception and became a religious philosopher and seeker.

For Burtt, psychology became a context for examination of much more than his personal life. He found ways to use it to consider universal questions, such as the possible integration of science and religion. Burtt's daughter, Freddie Burtt Brinster has had this insight.

> Religion and science were two magnetic poles, both exerting a powerful pull on him. I think his lifelong struggle was to somehow bring them together and integrate them. He was far too much a scholar and rigorous thinker to be seduced by religious dogmas. But he was also a deeply spiritual man and not about to concede that science had all the answers. His title, "The Metaphysical Foundations of Modern Science," implies, I think that his interest in metaphysics was deeper than his interest in science. Consider too that while at Columbia he was also getting a degree from Union Theological Seminary, that he became an ordained minister and that he sprang from a long line of Baptist ministers. Even though he rebelled against his father's religious conservatism there was still some of that missionary zeal in his blood.[21]

An extraordinary optimism had marked the *Humanist Manifesto* and persevered in the thinking of those who engineered it. By the mid 1940s Burtt had come to see the Humanists as hopelessly naive, blind to human foibles. They blocked out their own capacity to reason by turning away from public comment about World War II. Probably Burtt would not have been able to support them as long as he did, had he not had another motive. The Humanists gave him the freedom and the forum to tout psychoanalysis as a kind of religious rebirth. It is unlikely that anything like the articles which appeared in *The Humanist* would ever have been published in any of the other professional journals of the day, and this is what gives these papers so much importance, especially since Burtt's correspondence is lost.

In 1951, the same year that *The Humanist* had sponsored the symposium, a new edition of Burtt's textbook on religion, *Types of Religious Philosophy,* appeared. The new edition differed from the first edition (1939) in several important ways. For example, the chapter devoted to a discussion of Humanism in the first edition was abbreviated and incorporated into a chapter devoted to both Humanism and modernism. In this chapter, Burtt disavows faith in human reason on the basis of his definition of "mind" as an "emergent product" of the material, Darwinian evolution of life, "just as life itself is a historical emergent from physico-chemical structures." It is preserved merely because it gives to its possessors powers of adjustment to environing conditions which prove in practice superior to the best faculties of competing organisms."[22] Taking a purely behaviorist point of view, Burtt argued Freud's position.

> [Freudian psychologies] show that empirically the dominant force in human nature is not man's faculty of intelligence, shared, as he fondly supposes, with divinity, but a horde of subconscious drives, passionate impulses, and cantankerous emotions, only fitfully illuminated and weakly guided by gleams of reason. In fact, reason itself, under Freudian investigation, proves to be no separate faculty, uncontaminated by the less seemly energies of mind...Clarity of insight and impartial truth are at best

> ideals toward which *Homo Sapiens* may haltingly approximate; in actual practice his most objective thinking is permeated by irrational complexes and distorted by blind urges which constantly lead him astray. Reason is at best largely rationalization, and the operations of what we call 'understanding', especially in matters metaphysical and religious, are mainly wishful thinking...the real world in its cold and implacable unconcern for human interests is too intolerable for men to face, and therefore they have overlaid it with the fantasy of the warm projection of their hopes and aspirations.[23]

Was Burtt proposing that reason was just another evolutionary adaptation comparable to the categories he identified in the manuscript, *Genesis of Hypothesis*? Did he think of it as somehow incomplete as an emergent property? He certainly reaffirmed his view that man does not share any privileged position resonating with divinity through mathematical realism or any other intelligent scheme.

Another extraordinary aspect of the new edition of *Types of Religious Philosophy* is that Burtt responds point for point to Randall's criticism of him in *The Humanist* article back in 1947. John Randall, Jr., too, had lost the exuberant faith he had once had in human reason, back in the glory days of the 1920s, when he had written *An Introduction to Reflective Thinking* in collaboration with the Columbia Associates, and later, *The Making of the Modern Mind*, based on James Harvey Robinson's inspiration. But the reasons for Randall's circumspect point of view and Burtt's were different, and they were part of what had separated them in the 1940s. Randall thought people required a larger dose of symbolic religion; Burtt, thought psychoanalysis was the answer. The two men divided permanently at that time, their friendship over.

Burtt's writing around the period of his psychoanalysis has a different flavor than his earlier work. His vociferous confession of his spiritual rebirth and his more or less complete absorption in the philosophy of religion disappointed his wife, Mildred, alienated him from John Herman Randall, Jr., and cast a shadow over him at Cornell. He undertook a vigorous campaign to be understood and produced another major component of his original philosophy of mind, a method for open-mindedness. This was Co-operative Evaluation.[24]

Burtt wanted to see problems in philosophical method as problems in communication and recommended co-operation, tolerance and understanding. By ignoring new developments in logic during the 1940s and charting his own path instead, Burtt was making a conscious choice to become an outsider to the fashions of the day and to place himself outside philosophy as practiced at Cornell.

Although Burtt turned his attention toward religion and Eastern philosophy after the 1940s, he continued to think about the concept of presupposition as the foundation for the world's various idea sets. He always favored a critical, historical approach to epistemology while he continued to look to science for models describing standards of knowledge and truth. These themes were grounded in Burtt's all-consuming search for meaning and purpose in human existence, which he believed should crystallize into one most reasonable, most valid truth, based on consensus.

Finding a method to get at this truth became the core of his philosophizing after 1940. The quest for world synthesis became the cornerstone of his later philosophy, as he advocated his own method, co-operative evaluation, to be learned as part of the human journey.

In Search of Philosophic Understanding (1965) is the mature product of the method of co-operative evaluation. In that book, Burtt claimed that there were no "absolute presuppositions." He wanted to show how philosophers can prevent unconscious motivations from subverting the entire philosophical enterprise. Only if philosophers learn to examine their valuations as inclinations or preferences in critical discussion with other philosophers could total skepticism be avoided. For Burtt presuppositions are not fixed, but fluid, and rightly under the control of conscious individuals who confront them as intelligible, that is, capable of being logically interpreted and logically changed.[25]

According to John E. Smith, who reviewed *In Search of Philosophic Understanding*, Burtt had got a hold of a promising, bold, new outlook which could "rescue philosophy from the strife of systems and unproductive dualities."[26] Smith noted in his review that Burtt's valuable contributions to philosophy were presented tentatively and self-consciously. This "vice" is a virtue that Burtt actively cultivated, but his soft, rational voice did not compel much attention during the years that followed the publication of *In Search.*

A softer voice had replaced the daring, aggressive and playful one of an earlier time. The tumultuous transition can be observed in the two articles for the *Humanist* and in the revisions to *Types of Religious Philosophy.* Burtt had been contacted by the Humanists and found their project appealing, but possibly dangerous. He responded to them immediately and thoroughly in order to help guide their project onto as constructive and prudent a path as might be possible. Burtt did not break with the humanists as much as he became disaffected by their program, especially in light of their doctrinaire position on the mind-body question and their unwillingness to work out theistic semantics in scientific terms. As the various strains of humanism matured they moved away from religious concerns as such and toward agnostic or even atheistic ethics.

Since Burtt was mostly interested in working out an empirical or scientific theology, he was naturally disinterested, especially since psychoanalysis seemed to fill the role of ethical theology for him. His goal for philosophy or psychology was a scientifically acceptable, empirically based theory to explain the human need for and experience of the spirit.

As for the articles in *The Humanist* in 1945, these must be seen as atypical of Burtt. There is nothing comparable elsewhere. Burtt was under a great deal of pressure in his personal life in the early 1940s and dealt with those problems through psychoanalysis. In his personal life, it helped him immensely. Professionally, insofar as a professional philosopher and academic needs the respect of his intellectual peers, it hurt him immeasurably. These were years of separation from his wife, when marital discord could cost a man dearly in academic settings. There were intense career pressures of a different sort also--logical positivism was full blown at Cornell and Burtt, along with the other older philosophers, was seen by the new-comers, Max Black and Norman Malcolm and their students, as a has-been.

RIGHT THINKING AND CO-OPERATIVE EVALUATION

Burtt first presented his method of co-operative evaluation in the 1945 revision of his textbook on logic, *Right Thinking*, which was designed for a two semester introductory course. The original edition had been published in 1928 when Burtt had just begun his teaching career at the University of Chicago. It had been first intended to support a course similar to the one he had been teaching at Columbia before taking the position at Chicago, *Reflective Thinking*. The earlier edition, *Principles and Problems of Right Thinking* (1928), had been based primarily on the text *An Introduction to Reflective Thinking,* written by John H. Randall and the other Columbia Associates in Philosophy. Successful during the decade of the 1930s, *Right Thinking* had gone out of print in the early 1940s and Burtt revised it in order to tap into a wide student audience for his own theory of value and his newly conceived theory of philosophical method, cooperative evaluation.

Declining to "defend these theories [and] likewise the theory of logic presupposed in the rest of the book against philosophical objections,"[27] Burtt attempted to present his views impartially, but ended up taking a purely psychological point of view. The defense, which he acknowledged as obligatory in the textbook was not undertaken there, but showed up in journal articles on philosophical and theological method published in 1946 and 1947. Probably intended to model cooperative evaluation, the defenses relied heavily on syntheses of the work of R. G. Collingwood, Stephen Pepper, W.H. Sheldon, Felix S. Cohen, R. C. Lodge, and A. E. Murphy. They are marked by a tentativeness that is in stark contrast to Burtt's earlier work, especially *The Metaphysical Foundations* and the two early articles subsequent to it.

Right Thinking has four parts. In part I, the emphasis is on clarity. In Part II, certainty is the key. Part III discusses scientific reasoning, verification of hypotheses in an examination of truth standards. The section is remarkable in that Burtt has abandoned his earlier perspective on the history of science, as a parade of world-views, and accepted instead, progress in science as cumulative. He takes a thorough-going classic position on causality, universality of experienced phenomena and even adopts Randall's history of science over his own. By 1946, Burtt had turned his back on "scientific metaphysics" and the doctrine of categories entirely.

In Part IV Burtt puts forward his innovation in method, where "reason becomes our guide in the enterprise of democratic cooperation; it is the dynamic intermediary through which the ends sought by one individual are kept in equalitarian and hence impartial relation to the ends that seem good to other members of the community. Its opposite in this context is dictatorial authoritarianism--readiness to impose one's own ends upon one's fellows."[28]

American philosophy since the 1930s has surely not been characterized by gentlemen speaking thoughtfully from armchairs in dimly lit rooms. Since it was established as a professionalized discipline, jealous of the status accorded to science and even social sciences, philosophers have been a testy lot and it is not hard to see why a man like E.A. Burtt must have begun to doubt his fitness for the environment of discourse the new philosophers were creating. There is some evidence of Burtt's shaky status in this wolf-pack. Yet there is something admirable about his amiable persistence and continued eloquence in adverse circumstances.

Also significant, Burtt came up with another definition for the metaphysician. In *The Metaphysical Foundations* he had considered metaphysics to be the science of discovering the presuppositions lying at the foundation of an idea set, similar to R. G. Collingwood's metaphysics, developed in 1940. In the late 1920s, he had described metaphysics as a science of mental categories of cognition and perception. In the revised edition of *Right Thinking* Burtt makes the metaphysician the universal mediator among various philosophical positions and their peculiar interests, based on underlying values. In his scheme he sees Reality as consisting of formal relations, facts and values. He classifies values as either "homogeneous" or "heterogeneous".

Homogeneous values are derived from classical value theory and contain three different value perspectives, more or less pure. They are:
1.) An appreciation for scientific truth (facts); 2.) practical value or 3.) aesthetic goodness. All three fields of value have an established methodology

for assessment. Less clear are the values in the heterogeneous realm--law, religion and metaphysics. The primary business of metaphysics is to establish "agreement" with reality, where heterogeneous values overlap or intersect with homogeneous ones. "The metaphysician is trying to reach judgments which will have the value of agreement with whatever reality is relevant.

Reality itself, like fact, transcends the field of values, while agreement with reality, like truth, is a value...The difference between them is that truth, being simple agreement with fact, is a homogeneous value, while agreement with reality is a heterogeneous value. The metaphysician is necessary because "reality includes formal relations and the various kinds of values along with facts"[29] and the metaphysician will mediate between them. Burtt wants to reclaim the traditional "metaphysician's universe" as the "totality of everything that can be significantly talked about...it is the sum of all realities,...the scientist's order of fact, natural and social, the realm of formal structures, and the various fields of value."[30]

Burtt argues that problems in fields of special interest are solved independently by the formal logician, the empirical investigator, the ethical philosopher, or the theologian, each working alone according to his own appropriate methods. The role of the metaphysician would be "dealing with problems that arise from the interaction of these various fields with each other." When theology conflicts with science, for example or when art conflicts with morals. The new metaphysician will show "whether it is wisest to follow the guidance of criterion drawn from ethics,...esthetics,...law,...theology, or whether some synthesis of these is possible and desirable."

The problem was to find some ultimate criteria to be used in reconciling conflicts between these fields of human value. To accomplish this, idealism is preferred over naturalism. Relational theory is to be preferred over psychological theory when the former is understood as choosing "the most inclusive vision of social value"[31] and psychological theory is understood as individual freedom.

He defined psychological theory as "the view that finds the essence of good in individual happiness, conceived as implying a maximum of freedom to do what one likes" and relational theory as that "which finds the highest good in a pattern of interdependence by which individuals are functionally related to the social institutions of which they are members."[32]

The Naturalistic and Idealist schools of philosophy can serve as the raw material of good metaphysics if they generate a "more inclusive synthesis." "Naturalistic" schools are characterized by those which "in the formulation of their criteria, emphasize relevant considerations arising in the field of science as decisive, reinterpreting in terms of scientific truth, so far as they can, all considerations by which the theologian, the ethical theorist, the art critic, etc., find themselves confronted. The 'idealistic' schools are those whose criteria of reality reflect acceptance of some ultimate doctrine about values, fitting the realm of scientific truth by some technique of harmonization into the system thus constructed."[33] At the foundation of Burtt's method was a deep-seated faith in a "democratic process of unhampered growth" which he saw as reconciliation among competing theories in any fields and the development of some kind of acceptable "more inclusive synthesis".

Burtt set out two prerequisites for the application of cooperative evaluation. **First** is recognition that "people honestly differ in their ultimate convictions regarding values" and an earnestly undertaken commitment to "forswear all dictatorial dogmatism, all pretense of absolute assurance for his own view" and a willingness to "hold his convictions tentatively, subject to revision and correction."[34] **Second** is a prerequisite social condition, "an open arena for thought and discussion" and thorough-going commitment by each thinker to "play his full part in this social task." This will involve adoption of a "super value" of "transcendent worth--namely the value of continued growth in insight into desirable ends through wider mutual understanding and the constructive criticism that makes it possible."[35]

This is how the democratic process of unhampered growth will make it possible for participants to recognize the value in theories previously rejected and thereby to gain "pioneering insight into novel values that have thus far escaped all contending theories." To achieve a glimpse into reconciling synthesis, Burtt proposed five steps for democratic cooperative evaluation. These are:

> 1. Each participant adopts, impartially, all the theories which he proposes to reconcile, with one exception--those theories which are essentially uncooperative. Those must be rejected. Any such theories are identified by intolerance and the likelihood that those who hold them will forcibly suppress competing theories.
>
> 2. An impartial common language must be accepted so that value-concepts can be compared equitably. Burtt proposed that participants cooperatively develop "a system of concepts in terms of which conflicts of value-theory can be impartially stated, and which are sufficiently inclusive in scope so that all

problems felt to be real by any competent participant in the discussion can be formulated in the medium thus provided."

3. The new common language is used to analyze each theory. Each theory is to be broken up impartially into its details. The analysis will be formatted in tables which make it possible to see points of essential agreement, points of contention and points on which there is neither agreement or contention.

4. Where points of contradiction are identified, "one alternative will have to be rejected and one alone accepted. The selection is to be made based only upon "the specific consequences to which each of the contradictory alternatives leads in actual human practice, [assessed] in the light of the clearest and most inclusive vision of social value to which we can attain, under guidance of the democratic ideal."

5. The final step is a "harmonious reformulation" of the points of the two theories not considered contentious or in agreement, followed by synthesis with the points initially in agreement and with the surviving alternatives of those points initially in contention. Still no such synthesis will make any claim to finality.

"There will appear from time to time, situations in which the most rational harmonization of theories previously available will not seem entirely satisfactory. The problems that thus challenge thinkers will then be met by further detailed analysis, projection of new theories, critical comparison, and systematic reconciliation of the sort that has just been described. The process never ends, just as the process of seeking scientific explanation never ends. But each gain is a genuine gain compared with the view left behind."[36]

For Burtt controversy in philosophy was the bitter pill of its contemporary practice. Acceptance, mediation and synthesis were his special elixirs. He was a man out of step with most other philosophers who made a virtue out of contention. The ability for introspection Burtt learned from his psychoanalysis made an indelible mark on his intellectual style as well as his emotional well-being. His brilliant, analytical mind had been unleashed on itself and now he was compelled to turn over every conceivable objection, leaving his arguments tepid and sometimes unconvincing.

In his autobiographical article, *My Path To Philosophy*, written when he was eighty, he recognized the part his own personality had played in his philosophical view-point, noting the reticence which marked him and acknowledging it as a liability.

> I am aware today that...sensitivity reflected in large part a powerful force that was at work along with my competitive drive. That force was an intense longing for the approval of people around me and a haunting fear of their disapproval. In my contact with other thinkers, those feelings led me to avoid belligerent controversy and to seek genial accommodation with philosophers I came to know, either directly or through their writings. For a long time there was a good deal of tension between this longing and the urge for victory over whatever rivals appeared on the scene. But it seems to me now that the need for approval increasingly dominated the competitive urge and found definite expression in the way of philosophizing which had become characteristic by the 1930s. The guiding, though unconscious, maxim during that period, in my attitude toward my fellow philosophers, was: "I want you to agree with me; but if you don't, I'll have to find some way of agreeing with you. Were I willing to acquiesce in unresolved contradictions, the outcome of following this maxim would have had no value; I would have accepted the diverse proposals of other thinkers irrespective of the conflicts between them. Happily, this outcome was precluded by a respect for logical compatibility; I could not fully agree with everybody. The way out of the otherwise fruitless quandary was a growing realization that every sincerely championed philosophy has some constructive insight that ought not to be lost.[37]

"With a bit of facetiousness but also with a nub of truth," Burtt characterized his method.

> Through accidental good fortune, I managed to turn a serious weakness into a philosophical virtue....The promising path...became for me the path of responsive openness to every such insight and of a zest to see how the partial truths thus glimpsed might be brought together in a coherent, all-embracing perspective. I came to believe that all philosophers would do well to exemplify this openness as best they can, and that my special talent lay largely in suggesting ways in which elements that most thinkers seem content to leave separated might be brought together in a constructive unity. Many such envisioned syntheses proved deceptive, they provided no dependable illumination. But I believe some have been valuable.[38]

The valuable works Burtt referred to were his books on religion, the area he came to feel most adequate to address as he moved farther and farther away from his early interest, placing science and religion on an equal playing field of value in the consciousness of man. By the 1940s, he had abandoned his radical conception of science as argued in 1926 and in the theory of categories of perception in *Genesis of Hypothesis* in 1935. He tended to bury his original thinking within expositions of the ideas of others and to wear his philosophical method, which emphasized understanding, as a kind of armor.

His published writing turned toward religious synthesis, and occasionally, philosophical synthesis

After his retirement in 1960, Burtt's writing was again laced here and there, with the notion that science principles can make no claims for universality and that order in nature is something imposed there by the selection processes of the mind, conditioned by culture and personal valuation. He had formerly abandoned this point in *Right Thinking*, Part III, *Reasoning as Describing and Explaining Facts*. There he had discriminated between physical and biological science, which he saw as factual, and other areas of human endeavor, clearly subject to valuation, such as history, law, religion and art. [39]

But, he had not really abandoned those ideas. He resumed his interest in the logic of presupposition after reading Collingwood's work, *Essay on Metaphysics* (1940). Collingwood's more systematic expression of his own thinking re-opened avenues Burtt had previously explored and lately neglected. But they were there, at the ready, to be explored again, and during the early 1950s they began to show up in his teaching, and in his private writing.

CHAPTER 12

A RETURN TO THE IDEAS IN *THE METAPHYSICAL FOUNDATIONS*

BURTT AND R.G. COLLINGWOOD

David Hollinger has compared the extent and quality of attention Thomas Kuhn has attracted to the level of intelligent criticism won by R.G. Collingwood's *Idea of History* and Charles Beard's work on the history of the U.S. Constitution. Hollinger's calling Kuhn a "junior partner" of Collingwood[1] ignores the fact that Kuhn did not read Collingwood, he read Burtt. There is no mistaking the identity of Collingwood's concept of presupposition as the basis for his grand reformation of metaphysics as a science with Burtt's use of the Pragmatic method in *The Metaphysical Foundations,* calling the presuppositions at the foundation of Newton's science, Newton's metaphysics. So much alike are the two conceptions it is conceivable that Collingwood was influenced by Burtt's work, which had been published in England and widely read there just when Collingwood was on the receiving end of discouraging reviews of his first book, *Speculum Mentis*. Burtt strayed away from his own version of the logic of presupposition as a science of metaphysics after 1935, but after the mid-1950s, he returned to an emphasis on the concept.

The similarities between Burtt's concern with a philosophy based on identifying presuppositions in *The Metaphysical Foundations* and R.G. Collingwood's reformed metaphysics (*Essay On Metaphysics*, 1940) are so striking that Burtt had to take notice.

When Arthur Murphy had come to Cornell he began teaching a graduate seminar class in metaphysics from 1946 through 1951, at least. Many graduate students remember it well. One of the required texts was R. G. Collingwood's *Essay on Metaphysics*, but most of the course centered on the writings of Charles S. Peirce, the most analytic of pragmatist philosophers with the most appeal to the analytic types at Cornell.

One of the graduate students was Professor Marcus Singer who recalled that both Burtt and Murphy were teaching a seminar section in which Collingwood was discussed in 1949-50 or 1950-51. Singer had occasion to sit in on both seminars, but he was being advised by Murphy and much preferred Murphy's style to Burtt's.

Another Cornell graduate student was Professor David Braybrooke, now at the University of Texas at Austin and at Dalhousie University in Halifax, N.S. Braybrooke was one of six students who had taken Burtt's seminar. Braybrooke admired his teacher and preserved his own beautiful set of class notes as well as the typewritten hand-outs Burtt prepared for the group. Very likely, Burtt taught the section for only one year, the Fall term of 1950-1951 and there were just six students regularly enrolled. Singer, who visited the group at least once, made seven. Murphy's section considered Collingwood's metaphysics as a minor aspect, while Burtt's seminar concentrated entirely on Collingwood. One gets the impression that Murphy's was the larger section because so many students remember Murphy's class, but perhaps it is simply a matter of Murphy's having taught his seminar for several years.

Professor Braybrooke's notes indicate exactly how Burtt's conception of a revised metaphysics compared with Collingwood's and where the two men differed. The notes also demonstrate that Burtt had not abandoned his earlier philosophy from the 1930s after all, but by 1950, he had added into the mix his new ideas on philosophical method. This was Cooperative Evaluation, developed in his textbook *Right Thinking* as a method for reconciliation and synthesis among all the world's philosophies.

Burtt handed out a typewritten summary for the November 15, 1950 class. It begins:

> Presuppositions --Their Metaphysical and Epistemological Status
> Summarizing statement of the issues which have been emphasized in our discussions to date.
>
> A. Collingwood and Burtt agree in holding that a wise reconstruction of metaphysics will make it a systematic study of absolute presuppositions. Are they right?
>
> The main alternatives are:
>
> (1) The positivist rejection of metaphysics as meaningless.
>
> (2) The belief that some traditional conception of metaphysics--as a general theory of being or a general theory of knowledge-- is still valid.
>
> An important consideration in favor of the C and B position is that both these alternatives appear to have presuppositions. It would seem to be worth while to bring these presuppositions to light, to analyze them, compare them, explain them. As Collingwood shows, this enterprise has much in common

with what metaphysics has done in the past, though in other terminology.

The objections Burtt wanted to raise against Collingwood's concept of reformed metaphysics turn on the question of whether or not Collingwood's absolute presuppositions hold together logically or illogically. Whether or not the "strains" between consupponible presuppositions, identified by Collingwood as the driving force of history, can be guided.

Collingwood's point had been that an absolute presupposition is simply believed without any necessary substantiation and therefore can not be proved true or false, nor can the confidence a thinker has in it be shaken. It simply *Is*, as an Absolute idea held unquestioned and unquestionable by the culture or the individual. Collingwood reformed metaphysics from the study of Being qua Being to an historical study of absolute presuppositions and the propositions which are supported by them. History, properly conceived, according to him, is the study of the changes in absolute presuppositions held by a culture or an age which are reflected in the phases of history evident when that culture or age is studied. The change is history; the historian wants to ferret out the "strains" in the absolute presuppositions which lead to shifts in intellectual life, but he can not study them if they are static. Only as they change is their existence revealed as the fodder of history.

> *It is [a] false assumption that an historical 'phase'-- a civilization, a phase of scientific thought, a set of absolute presuppositions -- is a static thing, whose relations with others can be adequately studied by comparing them and noting resemblances and differences. The essential thing about historical 'phases' is that each of them gives place to another; not because one is violently destroyed by alien forces impinging on its fabric from without by war or from within by revolution, but because each of them while it lives is working at turning itself into the next.... The metaphysician's business, therefore, when he has identified several different constellations of absolute presuppositions, is not only to study their likenesses and unlikenesses but also to find out on what occasions and by what processes one of them has turned into another....*
>
> One phase changes into another because the first phase was in unstable equilibrium and had in itself the seeds of change, and indeed of that change.... [2]

Collingwood thought the changes occurred because of incongruities, which he called "strains" within the constellations of absolute presuppositions.

> Where there is no strain there is no history. A civilization does not work out its own details by a kind of static logic in which every detail exemplifies in its own way one and the same formula. It works itself out by a dynamic logic in which different and at first sight incompatible formulae somehow contrive a precarious co-existence; one dominant here, another there....[3]

The historian, who by Collingwood's theory is now a practicing metaphysician, will "expect the various presuppositions he is studying to be consupponible only under pressure, the constellation being subject to certain strains and kept together by dint of a certain compromise or mutual toleration..." When absolute presuppositions can not be supposed together anymore (are no longer consupponible), they break their uneasy truce and the action of history will take place, the shift in phase will occur.[4]

Burtt was not convinced that there are absolute presuppositions in Collingwood's sense. He argued that Collingwood's concept denied the possibility of logically studying changes in absolute presuppositions. Burtt wanted to argue that the absolute presuppositions are changeable in some pattern guided by logic and that the history of culture demonstrates this. The investment Burtt had in hanging on to the logical transition between the presuppositions of an age was that he wanted to make his own philosophical method, Cooperative Evaluation, the cornerstone of a new task assigned to philosophy, guiding culture in its transition from one world-view to another and identifying the best, most nearly true knowledge we possess, as that which is achieved by consensus. Collingwood made no pitch for a single best synthesis of all absolute presuppositions; his reform was a method for studying history. Burtt, on the other hand, hoped to guide a philosophical synthesis which would mimic the single best theory of nature. Burtt's metaphysics would propose the single best philosophy of man.

Burtt's method had encouraged synthesis by clearly defining termswhich escape mathematical handling. He missed the present-day debate over commensurability of terms, including mathematical terms, by a mere twenty years.

It is not fair to say that Burtt's viewpoint would have been strong enough to add much to the discussion about paradigm shifts and incommensurability of terms across theories because he did not satisfactorily demonstrate how rational transition in idea sets occurs over time or even that it has occurred historically. By insisting on rationality, while declaring that Collingwood denied it, he was really up to something else, anyway. He wanted to wrap Collingwood's ideas around his own, following a style he had adopted in the 1940s. In the spirit of synthesis, he would systematically sift through the most promising recent philosophical ideas tying in his own where they fit and always coming back to his philosophical method, co-operation.

He seems to have wanted to illustrate that through-out intellectual history absolute presuppositions are replaced in some logical, comprehensible way in order to give context in contemporary discussion to his new philosophical method, cooperative evaluation.

In his 1965 book on philosophical method, *In Search of Philosophic Understanding*, he says, "[Collingwood's primary distinction is between 'relative' and absolute' presuppositions...If the course of thought in the present book is sound, no absolute presuppositions in his sense can be admitted."[5] For Collingwood the absolute presuppositions support relative propositions. The absolute presuppositions are analogs to what was previously studied as Being. For Burtt, the presuppositions are always relative. For Collingwood the Absolute Presuppositions can not change without significant upset in history because they represent the ground of being when the ground of being is properly understood as mind thinking. In Burtt's view the presuppositions all change or can be changed through logical processes, which can be historically documented and used to guide human beings into an awareness of how reality is constructed.

Fundamentally, the distinction between Burtt and Collingwood rests on the two separate approaches they had taken on the question of God. Collingwood held out for one absolute presupposition, illogical, unchanging and unverifiable---the presupposition of Pure Being as a fact of history. Burtt, on the other hand, denied God as an absolute presupposition and chose instead to seek an empirical demonstration of "the divine" by making God into a metaphysical category in the mind, like time or space. The new category, once defined, would mediate between the unknowable reality of the Divine and human experience. It would contribute to human evolution, a fuller life and more complete human experience. For Collingwood the science of metaphysics, reformed, was not an empirical science. It was, rather, a science of the reality of mind thinking the thoughts of the past in the present and working out the intellectual knots of the past. Collingwood's conception made the intellectual history of man into reality with a capital R. Burtt's conception had its origin in empiricism, which as a Pragmatist, he understood as all of human experience. It made the fullest human experience into reality with a capital R. In 1969, Burtt gave his philosophy yet another name, the All-embracing Philosophy of Man.[6] Here he conceived Reality as "an inclusive view of life and the universe" recognizing the fundamental error of former philosophers, "that a final explanation of reality can be achieved."

For different reasons Burtt took issue with Stephen Pepper's *World Hypotheses,* a book he otherwise greatly admired, and recommended to his seminar students. Burtt took Pepper to task for ignoring the wider historical context while "proffer[ing] an explanation of the way the ideas of any philosophical school originate in common sense experience." Burtt criticized Pepper for failing to "point the way toward a world theory more adequate than any now available...and under the guidance of this hypothesis...a standard of metaphysical truth."[7]

Pepper had identified four "root metaphors" as the origin of metaphysical hypotheses at the foundation of the four great philosophical trends current in his day, realism, idealism, pragmatism and naturalistic mechanism. In another article, *The Problem of Philosophic Method* (1946), Burtt explains further how his opinion diverges from Pepper's. Burtt is looking for a single perspective, a "balanced," "superior" philosophical position, and he wanted to recommend certain criteria to identify that ideal. Pepper had settled on four irreconcilable root metaphors.

> Pepper holds that each of these four provides its own criterion of adequacy which we cannot at present satisfactorily transcend.[8]
>
> [Pepper's] "root metaphor" doctrine justifies...the rejection of all eclecticism and all attempts to synthesize theories drawn from different metaphors. [Quoting from Pepper] 'If world hypotheses are autonomous [as they must be if derived from different metaphors], they are mutually exclusive. A mixture of them, therefore, can only be confusing.'
>
> It is the unqualified rejection of synthesis of different philosophical systems as a promising procedure that constitutes in Professor [Wilmon H.] Sheldon's mind the most dubious feature of [Pepper's theory].[9]
> [Sheldon was also recommended reading.]

Taken in this context it can be seen where Burtt was going with his seminar group teaching his logical method, cooperative evaluation. Reading the hand-out notes in light of Burtt's opinion of Pepper helps to explain his contrary position on Collingwood:

> B. On the further basic question: What should such a systematic study consist of? There is a sharp divergence of opinion. [Between Burtt and Collingwood]
>
> (1) Is Collingwood right in holding that it should be simply a historical study of APs actually held by such and such people, at such and such a time, etc., supplemented by explanations of why they were held and by some comparison of different 'constellations' of AP's?
>
> His main reasons for holding this are:
>
> (a) Each constellation of AP's provides its own criterion of truth, reality, consistency, etc.

(b) There is no more ultimate criterion by which a constellation can be judged.

(c) While no constellation lasts forever, the process by which it is replaced by another is not a rational process.

(2) Is Burtt right in holding that over and above such a historical study, critical evaluation of different constellations of AP's is possible, leading to the emergence of a more impartial and inclusive constellation which is thereby shown to be more reasonable than the constellations it replaces?

His main reasons for holding this are:

(a) Such a process is actually going on historically, and is exemplified in every discussion that is more than an attempt of two people to convert each other.

(b) While no super criterion exists at the beginning of any instance of such a process, it emerges at the end and can be seen to have guided the process leading to it.

(c) When one constellation of AP's replaces another in this way, the process is a rational one. Its ruling canons (impartiality and inclusiveness) are rational, and rational techniques (e.g., "generic definition") can be devised to guide it more effectively. Such techniques have played a constructive role in the history of *philosophy*.

Burtt's reference to the rational technique, "generic definition" was an introduction to some new ideas he was then pursuing. An article in the *Philosophical Review, The Generic Definition of Philosophic Terms*, was published in 1953 and read before the Philosophical Club at Cornell in 1951. There he argues that philosophy is an agent of intellectual progress.[10] In what had become typical of him by then, Burtt is evaluating the ideas of others and applying his own philosophic method, cooperative evaluation, to what he sees as the significant contemporary philosophical problems. The paper had been delivered to the Philosophy Club at Cornell in November 1951, just about a year to the day after Burtt had prepared the typed notes for his students.

Burtt addressed the "linguistic responsibility" of philosophers to explain their own terms as well as to render the service of linguistic definition to the sciences. Noting that there are alternating periods in philosophy as there are in science, "a period of pioneering application of relatively new ideas being succeeded by a period seeking the systematic adjustment of ideas already in the field," Burtt's concern is for an era of synthesis, which he imagined the present era must be. He made the case for clarification of the language of conflicting schools of contemporary philosophy. He would "trace the basic issues between conflicting philosophic schools to their roots,...[and] find that those roots consist of diverging definitions, explicit or implicit, of certain categories or major concepts." The metaphysical categories identified as instruments of perceiving the physical world of science have their counterparts in philosophy. Just as in science, categories of philosophy correspond to the terms philosophers use as instruments in their branch of thinking. Burtt identified them as knowledge, relation, meaning, experience, reality, usage, logic (penciled in), and value (later scratched out). The wisest definition for each of these terms must be decided somehow in "light of contrasting concerns, with their associated presuppositions and achievements [in the field] when "partisan competitors" muddy debate. Burtt proposed rules for establishing a generic category in the event that no consensus is possible.[11]

In the case of trying to bring East and West together the problem turns on defining spirit, experience, and actuality for the Easterner and *dharma*, *karma*, *li*, and *Tao* for the Westerner.

With his idea of generic terms, Burtt anticipated one of the problems Kuhn's history and philosophy of science has raised, that science terms are not immediately commensurable across competing scientific theories. Burtt thinks that such terms become commensurable because scientists have got hold of a good method for dealing with them, mathematics. Beginning with an illustration from science, Aristotle's concept of causality as it changed throughout history, Burtt argued that a neutral language for philosophical terms must be adopted in the same way as he had imagined was already the case in exact science. Burtt equated objectivity with mathematical treatment. Of course, Kuhn's work resulted in the demonstration that science mathematical terms might not be as commensurable as Burtt believed they were. Perhaps the philosophers of science might look to Burtt's rules for cooperative evaluation yet for some help with their incommensurability problem and discussion.

Professor Braybrooke remembers that Burtt "cheerfully carried on with philosophy in his own style" in spite of the strongly analytic climate around him.

> I remember Burtt's sprightliness and enthusiasm in [the seminar on Collingwood's Metaphysics] class; and his repeatedly saying that we were engaged in a "pioneering investigation." That seemed a little incongruous, given the climate of philosophical opinion both at Cornell and at Oxford where Collingwood had stood back from the wave of pioneering in linguistic philosophy just as Burtt had stood back from it at Cornell, where in my time (1950-1952, after which I went to spend a year at Oxford and

> there wrote my dissertation for Cornell) Max Black and Norman Malcolm were the dominant figures...
>
> Burtt may have stood back; but he gave no sign of mortification and carried on cheerfully with philosophy in his own style....[12]

Burtt was not entirely without a philosophical community of his own. They just did not happen to be in fashion during the 1950s and not anywhere in evidence at Cornell. For the Collingwood seminar, Burtt drew on the work of several thinkers besides Stephen Pepper. One was W.H. Sheldon who had been at Columbia in Burtt's time and before. Sheldon's book, *America's Progressive Philosophy*, offered a conception of reality similar to Burtt's own idea.

> We shall ask, not what *ought* reality to mean to intellect, but what *does* it mean to the common man in his working life, and how does he come to believe in a real over against an imaginary world?[13]

Sheldon's general answer is experience, merging the idealist and realist viewpoints in action. In action, we experience the object as independent from ourselves at the moment of our action and in genuine contemplation the same is also true, although in the later case we have no way of proving it. For all of that, Sheldon takes the middle road along with Burtt and maintains that there is some truth in all philosophical views. None are truly adequate and some blended philosophy will prove a better explanation of what is.

> We take heart at man's liberation from one dogmatic limitation after another, and declare that there are no eternal a priori restrictions on reality, no universal criteria pinning it down to some one form rather than another....And so time honored hostilities of the schools have no ground; idealism and materialism, scholasticism and process, monism and pluralism may all be true together.[14]

Finally in a manner reminiscent of Burtt's favorite point Sheldon concluded:

> All the elements of reality are compatible with one another. This, if we realized it, would abolish the wasteful civil war within the philosophers' fold and free his energies for constructive discovery of nature's ways, and an increasingly fruitful plan for human life.[15]

Also suggested to the seminar students was Dorothy M. Emmet's *The Nature of Metaphysical Thinking*. Emmet was interested in pinning down just what kind of knowledge science knowledge is and had cited *Metaphysical Foundations* in her book. She thinks science knowledge suggests:

> Some sort of rapport between our intellectual processes and nature beyond them. But it may not go so far as an identity of structure. It looks rather as though certain structures within certain processes in nature may be projected in a systematically distorted way in our sensory experience and in the intellectual constructions based upon it. Distorted, since we are responding to the environing processes and not reproducing them; and the response is shaped by the selective emphasis given by the character of our own sensory and mental apparatus.[16]

Here is a restatement of some of Burtt's own ideas in The *Genesis of Hypothesis*. Apparently, he had not forgotten the manuscript and by assigning Emmet, he could resurrect some of his earlier grand program and bring his own thinking into the discussion.

Other thinkers on the seminar reading list were E.T. Owen, *The Illusion of Thought* and Albert Hofstadter, *A Conception of Empirical Metaphysics*. Hofstadter says that, "metaphysics is embedded in the mind of man, insofar as he understands anything. It is embedded in the mind of the scientist, and it must be in order for him to be a scientist...[This says that] just as soon as a man attempts to explain something, as soon as he attempts to order data in a cognitive way, there will be some form of order which he uses."[17] "Metaphysics should aim at developing a categorical theory, i.e.., a theory of types of order and types of entity *qua* functioning within such order, as well as a cosmology, i.e., a theory of world-order, in which it is shown how the types of entities combine to form a cosmos by virtue of their metaphysical characteristics."[18] This is what Burtt had wanted to do with *Genesis of Hypothesis*.

The article by Owen, which the seminar students read, is of a different nature. But, it, too, puts the reader in mind of ideas Burtt had advanced in the 1935 manuscript, *Genesis of Hypothesis* and in his review of Cohen's *Reason and Nature*. It opens the question, what kind of reality is inaccessible to thought? "Reflection creates the illusion that what thought sees is outside itself...When thought finds that its revelation of reality ultimately conceals it,...that it can give no final answer to its own question 'what is reality?' it should look for an explanation of its failure in the peculiarities of its own nature, and not conclude that it has discovered its own limitations....What thought sees is always and only things as they are in thought."[19] For Burtt this hardly implies that there is no reality beyond man's thought. On the contrary, it allowed him to open a new question with his students--what lies beyond thought? In seeking the answer he wanted to approach his own definition of reality, which would be the combination of all the world's great ideas into some super "all embracing philosophy of man".

Burtt was, indeed, "still very much pursuing philosophy in his own style." He responded to P. F. Strawson's *An Introduction to Logical Theory* (1952), which was an attempt to apply the principles of ordinary language philosophy to metaphysics. Burtt's article, *Descriptive Metaphysics* (1963), makes these main points in criticism of Strawson:

> Strawson is mistaken in believing that there is a changeless core of metaphysical categories, and also in believing that a metaphysical system can be purely descriptive, avoiding any revisionary taint....
> ...The conceptual system with which "we" are operating may be much more changing, relative, and culturally limited than Strawson assumes it to be. But it fills its role in a wider and more inclusive setting, in relation to which these limitations can be seen. Hence our thinking is not restricted to operations *within* that system; and the metaphysician cannot perform his distinctive task without taking into account the relations of that limited system to the dynamic whole of human communication which is not yet articulated in any definite linguistic form.[20]

Burtt wants Strawson to discuss the degree to which the metaphysical structure he wants to lay bare, through the study of language, is contingent or logically necessary, and how it compares to traditional logic or traditional *a priori* metaphysics. Burtt thought that he had found a method for finding the logical rules of structure our minds follow in constructing reality, describing how we adopt metaphysical categories. He summed up:

> For we surely can conceive of rules of thought that hold good for all human cultures; it is these that were sought in the long history of logic, and the actuality of such rules would seem to be presupposed in any impartial comparison of different languages with their varying categorical schemes.[21]

The necessary fields of research, the history of philosophy, religion, culture and now, language, on a world-wide scale are so vast that the task of identifying and documenting the categories for all the major world cultures will be a long time coming. Acknowledging how difficult it will be to formulate the rules, Burtt wants to suggest his own "principle of tolerance in understanding [which, he says] can...be as usefully invoked in the attempt to understand general logical and grammatical structures as in the analysis of specific concepts which is undertaken in...the philosophy of perception or the philosophy of mind."[22] The "limitations of our conceptual system" are not inevitable. Rather, we need to "learn to talk better in the presence of reality." Of course, Burtt is referring to his own method, the self-corrective, increasing critical awareness he had recommended in his last revision to *Right Thinking, A Study of its Principles and Methods* (1946), cooperative evaluation.

The decision to write out his own method and ignore "propositional or truth functional logic and [Max] Black's Critical Thinking"[23] had been another turning point. It launched Burtt clearly and completely on his own path. With that he set himself permanently at odds with the wave in philosophy at Cornell.

Burtt was interested in change. As we have seen he believed that change is the only reality we can apprehend and study, especially historically. He was a committed evolutionist. Burtt believed that human kind was moving forward, or should be, and this forward movement was something he wanted to guide himself and to teach others to guide as the basic reason for living. In the remainder of his notes for the *Metaphysics Seminar* he went on with some questions which he asked the students to address for themselves.

> (a) How do presuppositions originate? [Here he recommended that the students read Stephen Pepper's *World Hypotheses.*]
>
> (b) How are AP's related to their cultural background? (It seems clear that this background is <u>one</u> factor explaining why an individual holds some of the AP's that he does.)
>
> (c) When a constellation of AP's is replaced by another, is the process sometimes sudden or always gradual? And if the latter, does this mean that only a few AP's are changed at any given time, the others remaining constant?
> [Is there a "paradigm" shift or more gradual and logical change?]
>
> *(d) Under what conditions does the mere fact that one becomes aware of an AP throw doubt upon it? (It seems to be the case that sometimes awareness of an AP makes us skeptical about it, whereas sometimes--as with the AP of causality--awareness that we are holding it does not lead to doubt it.)*

In essence, Burtt was convinced that it is the mind of man which is logical or can be trained to be so. The universe is illogical. Our understanding of how our minds process experience and our increasing awareness of the peculiarities of other minds similarly engaged are the necessary first steps toward understanding anything outside of us. There is a single route to understanding ourselves. We must look at change, look at differences among ourselves; these are the only constants in our experience. History and culture give us the information from which we can extract the necessary raw material. We can not dependably rely on the external world for a picture of reality, but such a picture will become clearer if give up the search for certain truth and concentrate on the relative realities we create through mental constructions. Animosity and strife do not facilitate the evolutionary destiny of man, which is to come into a better relation with Reality, proper. Therefore, at this particular stage in human history some dependable, self-correcting method of communication must be invented and adopted. This will be similar to the scientific method, but it will concentrate on human reality rather than externals in nature. Just as the methods of science must assume, as conscious metaphysics, an orderly, intelligible universe, the methods of philosophy, or human understanding, must assume that minds are capable of open, honest communication (intelligibility) and that all people through all time have operated from the same kinds of logical mental structures (universality). Just as science looks for patterns in nature, communicated impartially by mathematics, so philosophy must look for logical patterns in the changes in world-views presented in studies of history and culture. The language of mathematics will not do for this endeavor, so a new common means of defining terms must be had. We must presuppose the logic of change in the same way science presupposes logical order in the universe, else we shall not make any progress.

Burtt had done ground-breaking research in the history of science, documenting that knowledge is not certain unless it is the knowledge of change. In this, he shares the opinion of R. G. Collingwood, who held that all facts are historical, and therefore subject to change in the mind of man, including the facts of science. Burtt had made some convincing demonstrations to illustrate that experience does not equate with physical reality, that imagination and the logical processes of mind interact with stimuli to construct relative realities.

After teaching the seminar on Collingwood Burtt's attention turned back to the concept of presupposition, a concept he had substantially abandoned in favor of developing a new philosophy of mind based on a metaphysics of categories of experience. Collingwood had said that metaphysics is the science of absolute presuppositions. He had written:

> People do not need to analyze their thoughts very deeply in order to find out that there are a good many things they take for granted without asking whether they are true; but his expression generally means not that they have decided on consideration that it would be nonsensical to ask whether they are true; but that they have asked this in a half-hearted way, and have been satisfied with answers that would not have satisfied a resolute and unprejudiced inquirer....But things are not quite so simple as that. To begin with, people may have a motive for deceiving themselves and each other. Where certain things which may happen in people's minds are conventionally regarded with disapproval, the lengths to which people in whose minds they actually do happen will go, in order to persuade themselves and others that they do not happen, are most remarkable....It is only by analysis that any one can ever come to know either that he is making any absolute presuppositions at all or what absolute presuppositions he is making.[24]

This is the concept, expressed in *The Metaphysical Foundations* as the metaphysics of an age, a culture, a personal mind-set or the metaphysics implied in a method. In the 1972 article, *My Path to Philosophy*, Burtt recalls his interest in presuppositions and notes that his interest "assumed an increasingly central role in my developing point of view" while he was teaching at Cornell.[25] It evolved in tandem with his concept of valuation, the part of human nature which favors some presuppositions over others.

> A presupposition is a hidden or tacit premise underlying any statement or question or piece of reasoning. And besides specific presuppositions implied by any particular statement there are basic presuppositions, revealed (when one looks for them) in all statements which express this or that way of thinking....
> More and more it seemed clear that the set of presuppositions which ground a given point of view always rests on some underlying valuation...[26]

Over the years Burtt came to adopt Collingwood's concept of presuppositions and make distinctions between basic and ultimate presuppositions himself on the same basis as Collingwood made the distinction between absolute presuppositions and relative propositions. The ultimate ones are expressed "in a whole cluster of ideas and even in the entire way of thinking characteristic of an individual or group."[27]

Why had Burtt concentrated on categories of experience as metaphysics rather than his concept of presupposition in the 1930s? One reason might be that Burtt was in the U.S., where empirical and analytical schools of philosophy carried the day. If there was any chance of being heard it was through the Pragmatic route, not the Idealist, neo-Kantian one, which Collingwood followed. In the U.S. it is action and outward experience which the culture values, not mind, not the inward experience of thought. Burtt was interested in the possibility of our apprehending a deeper sort of reality.

Burtt summed up *Philosophy, Yesterday, Today and Tomorrow* with what were to be his last words on the "philosophy of man." He wrote:

> The phrase "philosophy of man" is a short equivalent for a philosophy of the universe in which man fills an inescapable central place--inescapable because our concern is always with what he actively experiences as he interacts with other realities. Hence our pursuit of understanding is no longer a quest to find out what the world is, but instead what human beings are taking it to be or making it become.[28]

After this much was accomplished, the way would be opened to introduce the possibility of a reality beyond experience. Earlier in the same work he had written:

> The most general laws that human beings are capable of discovering are then not the broadest generalizations of physical science--as long as man keeps trying to explain his world and adopts new ways of doing so, they are laws of growth in experience and understanding; those are the all-inclusive laws.
>
> To recognize these general laws and pursue science accordingly would bring a wholesale transformation in some of our basic presuppositions. But surely it is a live possibility--even, given time, a likelihood. If that transformation can come about, radical changes in other basic presuppositions can come about too.
>
> So the ultimate kind of prediction that thinkers are challenged to make is not prediction of specific physical or social events, important though that often is, but a wise forecast as to how future thinkers will describe and explain the world. One who succeeds in this venture would live and think in a much broader time-span than is experienced otherwise; his or her universe is a vaster reality.[29]

As he immersed himself in the study of religion, psychology, and Eastern philosophy, Burtt came to believe that the rudiments of a philosophy of human inter-dependence and mutual responsibility for a world community were already present within the major world civilizations, waiting to be identified through historical studies such as the one he had completed for his doctoral degree at Columbia in 1923.[30] Burtt thought that any true philosophy would have to be an all inclusive, universal philosophy of synthesis, incorporating great ideas from all the world's civilizations. As his thought matured, Burtt's pragmatic turn of mind led him away from an emphasis on universal cosmologies as described by Newton and Einstein in the astronomical sense and toward the opinion that the cosmos most worthy of attention is the immediate one, the global community of man.[31]

Burtt admired Ghandi's teaching. He had traveled to India and had an opportunity to meet him in 1949. Burtt identified a "distinctly novel element" in Gandhi's thought, *Satyagraha,* which Burtt translated as "utter truthfulness". *Satyagraha* is truthfulness that is shown, "not primarily in ability to apply the method of science to the details of the physical world, but in an absolutely honest facing of oneself, and a full acceptance of whatever is thus disclosed." In this concept we can see Burtt's own experience with psychoanalysis and his own wrestling with psychological truth. What he identified in *The Metaphysical Foundations* as a man's unconscious metaphysics, the presuppositions upon which logical idea sets are based, was transformed in the later philosophy to include the presuppositions underlying our moral choices and philosophical preferences.[32] It is something like a Buddhist version of St Paul's definition of Christian love. St. Paul tells us we do not always know wherein lies the good we would do out of our "love" because we do not always love selflessly. Gandhi tells us that "the good always comes from following the path of truth...Searching honesty is always likely to uncover blind or selfish motives in one's pursuit of what appears good."

Where all of this leads is key, because it is the quintessential later E. A. Burtt, albeit far removed from the theory of cognition and perception described in *Genesis of Hypothesis* or the new philosophy of science suggested in *The Metaphysical Foundations*. Quoting Gandhi, Burtt wrote out what is essentially his own view. To follow *satyagraha* in the "pursuit of Truth requires a recognition that no one man, no one party, no one class, no one race has the whole Truth; ...all human views are partial."[33] This is obviously not relativism, nor is it theism. Actually, Burtt's point of view is an invitation to search for meaning in life by searching for the Truth. The difficult thing is that Burtt does not believe that The Truth can be ever really be found and yet full acceptance of this fact does not diminish the value of the search.

In a letter to his friend, Stuart Hackett, who had written a piece for a *Festschrift,*[34] planned by Samuel Lindley, Burtt commented on Hackett's essay by beginning "I have done a good deal of pondering on your paper for the *Festschrift*....Of course, I will continue to reflect on the fundamental issues raised, but I am probably as able now to sketch my reactions as I will be a few years hence." After dealing with theistic interpretation of God, which Burtt rejects, he treats the nature of Reality.

> The other major issue...concerns our logical presuppositions, and boils down to this. Does honest and responsible reasoning require us to assume that there are "absolute structural principles of reason?' This assumption used to appeal to me, but no longer does. You yourself acknowledge that any attempted formulation of those principles must be regarded as tentative, not absolute, and I presume you would acknowledge that the history of logical theories shows differences in the way those principles have been conceived at different periods and in different cultures. Surely, further differences will appear from time to time in the future, which can not now be predicted.
>
> Do not these considerations have a vital bearing on how we can wisely picture the goal of that historical process? I picture that goal as an ideal of dependable inference or reasoning, which can be approximated more and more but never attained. The conception of an absolute structure seems to me, in the light of these considerations, to be quite vacuous. What role does it fill, other than as an ideal of reliable reasoning?[35]

One way to look at Burtt's latter philosophy is to see it as having developed dialectically as a pragmatic philosophy of meaning achieved through the study of history, philosophy and religion. Properly studied,

Burtt's books could form a series and represent a complete historical outline of the facts upon which he formed his own philosophy and intended his readers to form theirs. *The Metaphysical Foundations* (1924), *Man Seeks the Divine: A Study of the History of Comparative Religions* (1957 and 1970), and *Philosophy: Yesterday, Today and Tomorrow* (unpublished but completed in 1987) can be taken together as historical studies with a pragmatic perspective. They cover the history of science, religion and philosophy, respectively. These books are pragmatic histories because they represent reflective studies of the processes of mind in the context of lived experience, both historical and contemporary. The American Pragmatists had intended to make history purposeful, to put meaning into present day events and ideologies by framing them historically within as complete a context of the past as any life time of study permits.

Burtt's book on method would be *In Search of Philosophic Understanding* (1965 and 1967). It is also a history of modern philosophy. Burtt had told his friends the book was "a foil against which readers can test their own evolving philosophies."[36] The history of philosophy is used to illustrate Burtt's own belief that all philosophies have important truths lying at their foundations; it is the task of the genuine philosophy to come to terms with the common truth in all the world's major idea systems, including science and religion.

By the time of his death in September of 1989, Burtt had completed eleven published books. In addition to the published books there were fourteen articles contributed to published books, more than twenty five journal articles, plus many book reviews.[37] Continuing to write well after his retirement from Cornell, Burtt left a minor legacy of unpublished manuscripts and privately published books, outlining a prophetic vision for the future of humanity. He was nearly 90 years old when he completed *The Human Journey*, based on the Stephanos Nirmalendu Gosh lectures he had delivered in Calcutta. Two years later he published privately *Light, Love and Life,* a book which is a small volume of personal reflections and guidance offered to the world's seekers among whom he numbered himself. At 94, he was still trying to publish the book, *Philosophy: Yesterday, Today and Tomorrow.*

It is not at all clear that Burtt was aware of the significance of his own work. In the preface to his unpublished work, *Philosophy: Yesterday, Today and Tomorrow.* He wrote:

> Sixty years ago I wrote a book on a hitherto rather neglected lesson in the history of Western thinking about the world. It was entitled: The Metaphysical Foundations of Modern Physical Science. I was concerned then for the many persons who were perplexed about the role of modern science and were looking around for what a historical understanding of its work might provide. They sensed that along with

> obvious achievements of science it has serious limitations which pose a challenge with all persons who are interested in the role of science. My contribution unexpectedly attracted considerable attention and is still finding new readers today.
>
> Since writing that book I have become increasingly frustrated at the failure of my philosophical colleagues to follow up this quest for wider historical understanding. Indeed, many seem ready to respond to the lure of parochial issues that give no scope to their philosophic genius. But the whole course of growing human experience lies before us with its challenge to our understanding. I am deeply distressed as I look at the continually missed opportunity to roam over the all-encompassing arena of the vast universe we inhabit and to pioneer boldly toward what grasp of its mystery is possible.
>
> In the past, religious seers have done what they could to meet this provocative need, but their doctrines no longer persuade intelligent minds. Magnificent visions must be possible that match their boundless outreach while freeing the insights they have glimpsed from all sectarian doctrines.
>
> So I offer this broad reconnoiter over the current scene. Its main common feature with that first book is the determined quest to understand both science and philosophy in the largest setting that might be relevant. The message is for all men and women who are interested in what philosophical thinking might offer to our growing life and experience.[38]

If Burtt is a pioneer for post modernism, his purpose was not to deny reality proper, but to open a new window on it. The legacy of his psychoanalysis and the dark period was that he permitted himself to follow his heart afterward. The raring-to-go radical Burtt, skeptical, critical from the pulpit, a "hard-headed" philosopher, who would never cower to anyone, neither his father nor Robert Hutchins at the University of Chicago, became, at last, a simple man of the spirit, whose final word on philosophy after a life time of thought and developing methods for philosophical understanding was simply love, after all.[39]

Burtt saw that the human condition allows only limited perception and cognition. We are able to see the world through a glass darkly--colored by motivation, valuation and emotion. He is asking what can we know? How do we know it? How do we know that we know it? The answers are not always reassuring because Burtt gives us no guarantee that we can know much beyond ourselves. All the while, his own commitment to the search for truth inspires us to continue to look for it along with him. If Burtt has been an unduly neglected thinker, it is because his work transcends the times. Still, it haunts the century like an uncomfortable ghost from an era too quickly eclipsed, and it echoes here and there in the work of contemporary thinkers, who, like Burtt, are not afraid of doubt.

CONCLUSION

Conflict and reconciliation played a role in Burtt's personal life as well as in his later philosophy. It influenced his decision to retire from college teaching in 1960 at the age of sixty-seven. Withdrawing from academia into the Quaker community in Ithaca, Burtt continued to write well into his nineties. His reputation among professional philosophers was diminished by a constellation of factors of no substantive consequence. One of them was the fact that he did not have much stomach for the aggressive nip-and-tuck, which has defined philosophy in America since the 1930s. Another reason Burtt's philosophy has suffered neglect is that he lived too modestly, and for too long. He made no attempt to stake a claim for his thought in posterity and given the fact that he was alive and well, no one made such a claim for him.

Burtt considered "peppery jousting" a vice,"slow, inefficient, and cantankerous compared with the kind of discussion that would replace it when philosophers are more aware of what they are doing and their motives."[1] When he retired from Cornell, as Professor emeritus, he placed no papers in the university archives, cultivated no former students to shepherd his ideas, and gave away the books from his extensive private library. He chose instead what he imagined to be unfettered philosophizing, pure and clean of politics, in order to cultivate the singularity and independence H. Floris Cohen has identified in him.

Disconnected from the university, he spoke and wrote outside of the professional community of philosophers, although he remained active in the American Philosophical Association and was named President in 1964. Between the time he retired from Cornell and his death in 1989, there were nearly thirty years. During all that time, Burtt's important work escaped the interest of philosophers of science and found its place among historians.

It may well be that he openly discouraged compilation and analysis of his work while he was living. According to one of his closest intellectual companions, Miss Peggy Walbridge, Burtt did not value a personal legacy and, in fact, burned an autobiography he had written as well as a life-time of correspondence. Although he kept in contact with important Western philosophers of his day, including Bertrand Russell, John Dewey, George Herbert Mead, Alfred North Whitehead, Aldous Huxley, and Eastern philosophers, Sarvepalli Radhakrishnan, Daisetz T. Suzuki, and Gandhi, America's younger generation of pragmatic thinkers, Sidney Hook, Brand

Blanshard, John H. Randall, Jr., as well as religious notables such as Reinhold Niebuhr and Paul Tillich, historian, Arnold Toynbee, the writer, Archibald MacLeish and many more, no trace of these letters remains.[2]

Burtt's second wife, Marjorie Murray Burtt, died in 1982 at the age of ninety-two. She had been a practicing psychotherapist right up until her death and confidentiality prescribed that her records be destroyed. Burtt built the fires himself near his meditation hut down by the rapids of Fall Creek, just below his home. Friends helped him carry the papers to the fires and watched the smoke and flames consume them as they listened to the roar of the water coming over the falls. Perhaps no one noticed that Burtt was burning his own papers along with hers.

Known as Ned, to his family and friends, Burtt lived out his later years in a serenity so remarkable that the power of its other-worldly quality has lingered in the minds of those who were part of his inner circle, mostly members of the Ithaca Society of Friends. The effect that both Ned and Marjorie Burtt seem to have had on some people is of near-mythic proportion. As regular hosts and leaders of Quaker meetings, they were openly regarded with a combination of love and respect. They were true and loving partners, generous with all the resources they had, material and spiritual, living out the last of their lives where they had begun their life together at 227 Willard Way, just off the green, above the Cornell campus. Unbelievable though it may be in modern times, the word "saintly" pops up in conversations remembering this pair.

In 1971 a *Festschrift*, for Burtt's eightieth birthday was in the works. Stuart Hackett, whose company had published *In Search of Philosophic Understanding* (1965 and 1980) contributed an article,[3] apparently a commentary on that book. But the project was scrapped by the organizer, Samuel Lindley of Honolulu, Hawaii. Lindley had been a graduate student at Cornell (1946-1949), interested in the history of religion with Burtt as his major professor. The articles were returned to the authors and the volume was not published. Instead, a tribute was organized and appeared in the journal, *Philosophy East and West* (October, 1972).

1974 was the last time that Burtt's work appeared in mainstream American scholarship.[4] At that time Burtt was eighty-two and seems to have struck up a congenial association with Peter A. Bertocci. Bertocci is best known for developing varieties of Edgar Brightman's (1884-1953) ideas, which have become identified with the philosophical school known as Personalism. Bertocci was Brightman's successor as Borden Parker Brown Professor of Philosophy at Boston University. Acting as editor of a collection of modern philosophers' ideas, *Mid-Twentieth Century American Philosophy: Personal Statements,* Bertocci included a reflection written by Burtt. In the essay, *Toward a Philosophy of Philosophy,* Burtt recommends that philosophers adopt the scientists' practice of using a communal effort to advance their work.[5]

> What kind of enterprise would philosophy become if it were generally assumed that the normal thing is not for a philosopher to round off his system and then defend it against all critics, but to keep growing--to hold even his basic ideas as revisable in the light of new insight which might at any time be gained?...Militant polemics would disappear; they would be replaced by sensitive collaboration between philosophers. [6]

In recommending that philosophy become scientific, Burtt had an outcome in mind different from what analytical philosophers were trying to achieve, which was to make the language of science more precise in order to more accurately evaluate the "certain truths" regarding the nature of the universe. Burtt noted instead that

> [t]he aim of establishing final truth, which dominated pre-modern science, has had to be abandoned....Science is now viewed, not as a quest to uncover the changeless structure of reality, but as the endless process of revising and improving what at present passes for knowledge....Philosophy thus pursued would be mainly concerned with basic presuppositions and values as they take form from time to time in this or that limited group...[7]

Burtt had begun his personal reflection for Bertocci's collection with a reference to *The Metaphysical Foundations of Modern Physical Science*. He wanted to argue that the lessons to be learned from that fifty-year old book pertained to developing "a philosophy of man." The following year Bertocci contributed an article about Burtt's philosophy to the journal *Idealistic Studies, Love and Reality In E.A Burtt's Philosophy: A Personalistic Critique.*[8] Bertocci chose Burtt's *In Search of Philosophic Understanding* as his point of departure, especially Chapter IX, "Love, Creativity, and Reality." Bertocci wants to clarify what he thinks is the relationship of persons to reality by exploring Burtt's study of love as an action of understanding. Burtt argues that this "action" is a thinking action, which can change what Burtt calls "the pliable and social nature of reality."

If Burtt was a little far-out using a word like "love" in a philosophical study, he had come by it rightly after a thorough-going definition. Bertocci sums up with the essence of what E.A. Burtt had come to represent to those who knew him in his last years, the "cosmic Person," embracing human understanding, orderly activity, well-reasoned, loving behavior, and most of all creativity. "Love and Reality" is what Bertocci chose to see in Burtt's philosophy. If Burtt was at peace with that, it testifies to who he had become at the end of a century of examining his own presuppositions. He had written to his friend, Helen Schantz, in November of 1981 on the relationship between love and truth. "I don't think that truth and love can be interchanged, but truth without love is not really truth." In January 1982, he continued the theme. "From time to time I have been tempted to overrate the logical clarification, and to lose the assurance that love and God are two names for the same Being..."[9]

Since his retirement, Burtt had been moving progressively further and further away from the mainstream philosophy, a place he barely belonged after 1935 anyway. Certainly after 1950, he became more and more a purely spiritual man. As such, he was out of step with the university philosophy departments across the country, concerned as they were with technical arguments about logic and language. Burtt is acknowledged among American philosophers, not so much for his philosophy of science or even his pragmatic insights, but for his having improved East-West understanding.[10]

In 1987, Burtt attempted to have his last major work published. The reviewer for Routledge & Kegan Paul, Ltd., Stratford Caldecott, did not offer much support for the manuscript Burtt had titled, *Philosophy: Yesterday, Today and Tomorrow.* He counseled Burtt to retitle the book, *Foundations of Philosophical Agreement.* The title chosen by the reviewer illustrates the book's content fairly well. In the early pages, Burtt once again recalled *The Metaphysical Foundations* and noted the continuing importance of the philosophical method he had demonstrated there. Caldecott's rejection letter was handwritten and contained these lines:

> In theory, this is an excellent way to expose one's basic presuppositions. However, it is so challenging that I cannot imagine it being widely adopted in today's world. The M.S. would have benefited, I felt, with some more engagement with other contemporary philosophers...[11]

In handwriting, feeble with age, Burtt listed Caldecott's criticisms; he was at this time already ninety-five years old. The elder Burtt who emerges from the latter period and from living memory is not the vintage Burtt of the 1920s through the 1950s. He was no longer the pioneer in the history of science, no more the pioneer opening up the East; he had become a pioneer for human understanding and a pioneer into the inevitable downside of the human condition, old age and death.

As an elderly man Burtt is a treasure with much to say and much to teach of wisdom, of peace, and of the inner quiet. But for all of that, it is the younger, vigorous Burtt, still wrestling and jostling with life's big questions, however unhappily, that interests us and instructs us best in our own wrestling, and it is there that contemporary interest is likely to lie for the foreseeable future.

Some of Burtt's highly respected books achieved classic status his lifetime.[12] Typical of things Burttian, these books are a bit of an anomaly. They are mostly textbooks, intended for undergraduate college courses, but in their pages Burtt defined and developed his unique philosophical method. Uncritically accepted as they have mostly been, Burtt's books on philosophy and religion are taken as points of departure and reference, but have inspired little commentary or scholarship themselves. Never has there been a seminal thinker, who seeped so quietly into the mainstream of American thought. Having liberally sprinkled his pearls across textbooks read by students, Burtt has been a little bit like a new-age Socrates, teaching an enlightened skepticism to generations of the best educated youth through-out Western Europe, the United States and Canada.

In the last thirty of his nintey-seven years, Burtt made very little effort to take credit for his original philosophy or its insights. Following the Buddhist principles he had adopted, he was unassuming and simple in his demeanor. Some thought he was shy. He was certainly modest and humble. People who had known him for years as a friend were startled to learn "Ned" whose company they so enjoyed was, in fact, E.A. Burtt, whose books they so admired.[13]

If Edwin Arthur Burtt is an unduly neglected American philosopher, this is a situation likely to change since the neo-Pragmatists and contextualist philosophers of science are gaining some attention in a new conversation about context-dependent reality. A major influence on this trend has been Thomas Kuhn's *The Structure of Scientific Revolutions* (1962) and the restoration of the status of Pragmatisim accomplished primarily by Richard Rorty's *Philosophy and the Mirror of Nature* (1979). Both of these authors have cited Burtt's *Metaphysical Foundations*.[14]

What Burtt can contribute to the postmodern conversation is balance. Among all the pragmatist philosophers, Burtt is the most accessible. The fact that he wrote specifically about the history of science as key to developing a new philosophy of mind is the most relevant because he intended to shake the very foundation of the modern mind. To read E. A, Burtt is to find a constant and continuing faith in the pursuit of knowledge while acknowledging human contextuality. To read E. A. Burtt is to be carefully and consistently schooled in the value of questioning presuppositions in order to achieve self-knowledge and mutual understanding with others. Three themes pervade Burtt's philosophy, co-operation and mutual understanding, questioning presuppositions, and acknowledging the importance of human values in rational thought.

E.A. Burtt has had a role in the history of the development of post modernism because the continued popularity of *The Metaphysical Foundations* has played a part in the development of the postmodern conversation, and the Pragmatic Naturalism at the foundation of the work continues to play devil's advocate with modern empirical science. Naturalists are committed to an idea of nature, which includes the inner worlds of the mind as well as the material world. Still, Naturalists never abandoned the possibility of real knowledge. Naturalists are in the tradition of modernity and of modern science, to be sure, but pragmatic naturalists understand natural objects and processes in terms of their functions and purposes rather than as fixed and finished realities.

Naturalists are suspicious of the idea of the mind as a mirror, faithfully reflecting a wholly independent reality and in this, naturalism has glimpsed many of the postmodern insights, such as acknowledging human contextuality in everything known. What sets E.A. Burtt apart from the other naturalist philosophers of his day was his fearless confrontation with modern science. His comprehensive historical study leaves little doubt that science knowledge has a human context, whereas, the other naturalists wrote almost exclusively on ethical theory, social philosophy, aesthetics, and religion.

Burtt was committed to the theory of experience. For the Naturalists experience was not just the accumulation of sense data, but an all-encompassing range of what people do within the world and undergo within their minds. Naturalism seeks to include all that can be experienced in the whole human range and to study and understand all aspects of human experience as equivalent to any other aspects of nature. In other words, naturalism includes man and his experience within the scientific study of nature.

In common with empiricism, naturalism takes empirical observation very seriously. Naturalism values the scientific method while it avoids the extreme rationalism of idealist thinking. What sets American naturalism apart from modern science is the value it places on complete experience as knowledge. This was especially true of the followers of John Dewey and E.A. Burtt. The naturalists wanted to incorporate creative expression, emotional experience, valuation and active manipulation into a complete interpretation of experience. As Naturalists, these thinkers ruled out the supernatural, and what this actually meant was anything that might reasonably be called traditional religion. This is because traditional religion is almost always associated with the idea that beyond our world of experience there is something else. Beyond nature and natural understanding, there is God or some such equivalent which is the creator and sustainer of nature.

In this respect, naturalism is at odds with religion and here is where Burtt diverged with the majority, hedging on the question of God as not empirically demonstrable, while he earnestly maintained the existence of human spirituality and the human need to believe in something beyond nature. But how do human beings experience the Divine? For Burtt experience is key. This is a question with which he always wrestled and one which he did not put to rest until the very end of his life. During some of his most productive periods he seems to have believed that the power of human understanding or awareness is the route to the spirit.

He questioned the universal application of rationality as something not necessarily evolved as part of standard human operating equipment, but something which would emerge, if guided by philosophers. Existential in orientation, he believed in "man's capacity to leave behind the limitations of his present experience" but he rejected Hegelian logical consistency as abstract, bringing him more in line with Kierkegaard's idea of "intuitive realization that only takes place in the heart and mind of the individual." His later philosophical program centered on the Kierkegaardian ideal of reconstructing "one's entire self, including one's ways of reasoning."[15]

The success of *The Metaphysical Foundations of Modern Physical Science* was not matched by Burtt's later philosophical work. The reasons for this are complicated and not even altogether clear, but the strong-hold of analytical philosophy at Cornell, where he was Susan Linn Sage Professor from 1932 until his retirement in 1960, is a major part of the story. His interest in religion, which earned him the reputation of being not just tender-minded, but intellectually soft, is a second factor. Paradoxically, the most engaging part of Burtt's life as a philosopher is probably part of what kept him from being better recognized by the professional community. This was his emphasis on action as the duty of a philosopher whose job it is to wrestle with the practical problems of the human condition rather than to write arduous proofs of logical schemes. Such a point of view was a hold-over from the by-gone age of William James and John Dewey.

In some ways, Burtt suffered the defects of his virtues. Having achieved some fundamental insight, both new and important, he over-generalized it. As any pragmatic philosopher must, if he is to be intellectually honest, Burtt came face to face with the question the Pragmatists have traditionally left unanswered. Is the only order in the world what the inquiring mind puts there? Like the earliest Pragmatists of the Golden Age Burtt placed such great stress on the active role of ideas in giving pattern, order and meaning to human experience that he underestimated the importance of finding the solid premises from which human experience can be interpreted and upon which it can be grounded. In order to be intelligible and attractive to the majority of seekers, some of these premises must refer dependably to some order of the facts in the world and not just to human habits of inference.

For Burtt there were two possibilities, neither one satisfactory--an order "grounded in metaphysical principles that are supposedly eternal" or a "network of causal and functional relations so conceived that thinkers can predict the future effects of forces...and can make them serve desirable ends wherever the causal conditions necessary can be controlled." The metaphysical order poses as absolute truth, but has not any explanation in terms of predictive order. The scientific order, imposed by modern minds, has provided only relative order, constantly subject to revision. "Science...now viewed, is not a quest to uncover the changeless structure of reality, but as the endless process of revision and improving what at present passes for knowledge."[16] After a lifetime of trying to reconcile science and religion, Burtt found no reconciliation but instead came to insist that all searches for system and order, all pictures of "reality" will be colored by the values operating in the personality of the individual thinker. Burtt's idea of a philosophy of man would be "mainly concerned with his basic presuppositions and values." Philosophy could guide mankind as a whole by paying attention to the history of ideas. "Metaphysics [would become] a search for the most general features that can be delineated...Ethics [would become] a clarification of the most dependable and widely shareable good for man that can be envisioned. Logic [would become] a quest for the most trustworthy pattern of inference from any set of premises to a conclusion." Always the philosopher would consider the "hard won" contribution of past thought, while he would always be open to its revision.[17]

In short, E. A. Burtt preferred to philosophize in the grand old tradition of philosophy as wisdom and to practice philosophy, which, for him, meant to guide individuals in their daily struggles with significant questions, and in-so-far-as it was possible, to guide the civilization toward a greater understanding of what the human condition can be. As his life wore on, he simply became more interested in understanding, on the deepest level, what went on in the minds of others than he was in putting forward his own view.

There is clarity and simplicity in Burtt's writing which belies the depth of the questions he was addressing. A cornerstone of his philosophical interest was the quest for a reconciliation between modern scientific thinking, with its emphasis on Cartesian duality and mathematical logic and the other dimensions of human rational experience, usually defined as philosophy and religion. Here, he shared the opinion of all the most adventurous minds of the Golden Age. The piecemeal, reductionist empirical methods of the sciences would not do in providing for a philosophy of man. Burtt wanted to discover some new epistemological model, appropriate to the way human minds work, to guide inquiry and to help decide moral choice.

He was critical of logical positivism in the early 1920s, long before his contemporaries came to similar realizations in the late 1930s and 1940s, because that philosophy tended to argue away central aspects of human thinking such as values, ethics and religion. He was convinced that an extensive historical analysis of most contemporary idea systems would reveal them all to be little more than "objectifications of the mood of an age, fitful and temporary."

A presupposition of his thought was that theological dogmatism and the psychological need for certainty in a hostile environment were impediments to the formulation of a satisfactory philosophy of mind. In these views he is similar to John Dijksterhuis,, whose influence he acknowledged eventually.[18]

After World War II Burtt expanded his interest in problems of knowledge to include problems of world understanding. For Burtt, any new scientific cosmology, as well as our hopes for global understanding and world peace, depend upon the successful formulation of a new epistemology. In this he came close to the early idealists who believed that all clashes of culture and social interests could be resolved in a "higher synthesis," representing the best views of all parties. He believed most of all in "understanding," which he expected to result in "good synthesis," one which would be accepted by all the diverse groups when misunderstandings were put away.

He worked hardest at understanding diverse philosophies himself and his achievements there are amply reflected in his textbooks. Ultimately he came to believe that the rudiments of a philosophy of human inter-dependence and mutual responsibility for the world community were already present within the major world civilizations, waiting to be identified through historical studies and philosophical understanding. In 1964, Burtt addressed the sixty-first Annual Meeting of the Eastern Division of the American Philosophical Association at Boston, Massachusetts, as its president. His talk was titled, *Truth, Understanding and Philosophy*. Burtt outlined why he thought that speculative philosophy had fallen out of favor because its truth claims were not demonstrable. He argued for the substitution of "understanding" for "truth" as "wisdom that includes knowledge but is not knowledge" in the scientific sense. Philosophic "understanding" would assume the role filled in the past by speculative philosophy and recognize "the inevitable relativity of knowledge in a dynamic universe--the fact that more adequate knowledge than we now possess always lies ahead and that wiser conceptions of what knowledge is likewise lies ahead."[19]

Burtt thought that any true philosophy would have to be an all inclusive, universal philosophy of synthesis, incorporating great ideas from all the world's civilizations. True is a troublesome word in Burtt's world-view because he is clearly a relativist, at least on the face of it. All the same, he seems to have held that there was such a thing as philosophical truth and this comes closest to his notion of "understanding." The front piece of his last published book, *In Search of Philosophic Understanding*, is inscribed "In the long run truth wins the race with falsehood and error--and it wins because it is truth." What this would appear to mean is that falsehood and error are just incomplete understanding; truth is perfect understanding, but we are not likely to achieve it. We come closest when we attempt to understand the other fellow.

Perennially optimistic in spite of the frightening political developments of the twentieth century, Burtt sometimes spoke as if real philosophic understanding were not so much outside our reach intellectually, as it is a challenge to our will. We have to force ourselves to be open and to acknowledge our biases. His philosophical methods recommended that the power of introspection and self-discipline should be brought to bear to counteract parochialism and to inspire us with courage to live in an uncertain world. In Burtt's philosophy "reality" is a much more complex affair than previous philosophies generally allowed, because he thought that unless fundamental motives and other human factors are understood to be part of our analysis and apprehension of reality, we will under-appreciate it even as it stares us in the face. Like the other naturalist pragmatist thinkers, Burtt thought that reality was individual experience within a specific context, but for him there was more. For Burtt reality expands through the effort of human understanding. Reality in its widest and most absolute sense is human reality on the broadest scale. It is accessible to a single individual in measures according to a calculus of effort in the patient application of the methods of philosophy, history and psychology.

As a young man, at the start of a university career, Burtt hoped that he would work out a philosophical system of his own which he would then spend a lifetime teaching and defending against anxious rivals.[20] He began with the most fundamental question in philosophy. Is there an objective physical reality apart from human consciousness of it and can this reality be known apart from a human scientific construction of it? Ultimately, Burtt did settle on an answer, which appears below. It is not taken from *The Metaphysical Foundations*, however, but from an eight page mimeographed document found among his papers and books, parceled out from his study after his death and stored on bookshelves at the landing of his English Tutor style house on Willard Way in Ithaca, New York. There is no date on the paper, but it appears to encompass all the lineaments of his lifetime of thought. This assessment of mine plus his use of the phrase "cybernetic automation" on the third page suggests the paper is one of Burtt's final attempts to pull his metaphysical ideas together.

> Existence in its infinite fulness (sic) is unknowable and has no value or significance of any kind except as the focus of attention, perception and theory select from it those data which can be ordered into a system permitting far ranging inference and evaluation. In this way a universe is created, a home, so to speak, for the valuing consciousness of man indefinitely expanding, by creative communication from person to person, culture to culture, and age to age throughout human history.
>
> The mind is not free to select and order data as it wishes. The data are truly in existence and must be found where and when and as they truly are. But the fulness (sic) of existence is always more than these data selected out of it. This we know because new theories, new interests, and diverse ways of ordering and focusing attention can discover data previously inaccessible to conscious awareness. Different ages, cultures and persons live in different worlds. All this indicates that existence is an inexhaustible (fullness) out of which many different worlds can be created. But none of these worlds can be identified with existence as it is prior to the selections and orders achieved by theories directing focus of attention to find data to fit the theories.[21]

By the time Burtt wrote the eight page manuscript quoted here he had extended his notion of different scientific world-views, passing in a parade through history, to include "many different worlds" not necessarily corresponding to the physical world as it is in itself, although he acknowledges that that world truly exists. The different worlds are culturally or socially conditioned. With this type of thinking Burtt has entered some of the most controversial arguments deriving from Thomas Kuhn's *Structure of Scientific Revolutions*. He has come down squarely on the side of the constructivists and opened problems of commensurability across scientific theories and more generally across world-views. For Kuhn radical shifts in scientific beliefs, or "scientific revolutions," result in new theories about the natural world, but the new theories are not necessarily commensurable" with the old. Old terms, even mathematically related terms, adopted wholesale into new theories are suspect and can not be assumed to have the same meaning as before since the old theory has been transformed without a standard for correspondence in the new. This is why, for Kuhn, science is not cumulative in the sense that a new theory simply corrects the errors and impediments to solving the problems insoluble in the older theory. Kuhn's position is that it makes no sense to talk about the world as it is because we have no way of checking our theories against it. No philosophy of science can test a match between scientific theories and absolute reality because we do not have access to a perfect description of that reality against which to check the theories. We have only the scientific theories and a scientific method that allows those theories to be tested in terms of themselves. Burtt shares Kuhn's view and goes further; he has extended it to include all aspects of human knowing.

Just as there have been problems with Kuhn's philosophy, there are some for Burtt's since one of Burtt's radical notions was to challenge the assumption that the universe is an orderly place. Order in the universe is fundamental to the scientific world-view; without it scientific principles topple and commensurability is impossible. If the possibility of various world-views or phenomenally experienced worlds is admitted then how is communication between people ever going to be possible? This is one of the same questions Kuhn's philosophy of science has raised. Troubling issues of philosophic understanding were a recurring theme in Burtt's thinking, too, but he did not turn to learning theories for answers, as Kuhn has done, rather Burtt looked to psychology and his own philosophic method, "cooperative evaluation."

Burtt's early work tested the essential modern scientific premise of universal order, the so-called laws of nature, when he placed their foundation in metaphysics. To do this he borrowed heavily from Hume's arguments, undermining the assumption that causality, once established for a particular phenomenon, can be extended universally to like phenomena. He ignored Kant's explanation, making causality a category of mental operation, and critically explored Newton's laws of reasoning to expose the ordered universe as a modern presupposition only. Hume's opposition to the seventeenth-century design argument became Burt's search-light. He used it to reveal the logical consequences of removing God from the cosmos. Those consequences are that another power must be substituted to design the order assumed to govern the universe and hold it together. But what could it be? This power is not empirically verifiable by any means we have found so far. According to Burtt, Newton, himself, was perfectly aware of this and "feigned no hypotheses" about it. For Newton, it was God who held the order of the universe in His hands.

Fifty years after The *Metaphysical Foundations* Burtt described his point of view then as "semi-Hegelian idealism...harmonized with the liberal Protestant orientation to which my religious searching had then led."[22] By then, he said, he had given up on the idea of philosophic systems altogether, but his methods of inquiry remained unchanged. This is the method of looking to the underlying presuppositions to penetrate to the core of any seriously proposed philosophy.

> I was more and more fascinated by the challenge to understand thinkers whose ideas reflect quite different ideological, religious, cultural, or historical backgrounds, and to seek a dependable method of bringing their positive insights together in a coherent whole.

> The crucial consequence of adopting this approach to philosophy was that before long I realized that by the route thus followed I would never reach a philosophical system such as I had been eager to reach...More insight always lay ahead, so it would obviously be impossible to erect a final philosophy.[23]

Although Burtt did not put forward any formulaic philosophical system in *The Metaphysical Foundations* his method is penetrating because of the way he used it to describe and criticize early modern science. In an indirect way he has made important contributions to the post modern notion that "reality" is more-or-less a mental construct. Burtt's work is a significant thread in the history of twentieth-century thought, going against the grain of the dominating trends in formal philosophy for most of the period. Those schools have emphasized empirical verifiability and used well-defined rules of logic in order to achieve a measure for truth. Burtt's persistence with ideas some might have considered out-dated, can now be seen as iconoclastic or even prophetic. In any case, he helped to save a place for speculative philosophy, which is now, in the final decade of the century, starting to find its place in the mainstream again.

Edwin Arthur Burtt's faithfulness to closely held philosophical convictions and to his own temperament led to ideological conflicts with his analytically-minded colleagues at the Sage School of Philosophy at Cornell. Analytic philosophy, imported mostly from Europe, but also homegrown as the "new realism," eclipsed pragmatic skeptical philosophy through the 1930s and after. The error in this one-sidedness has been addressed in the last decade of this century and this is part of what makes Burtt attractive to thinkers again. Now he seems to be postmodern and a good anodyne to the ills of modernism and of postmodernism taken to extreme.

NOTES TO
E. A. BURTT: *HISTORIAN AND PHILOSOPHER*

Notes to the Introduction

1 *The Metaphysical Foundations of Modern Physical Science: A Historical and Critical Essay,* (London: Kegan Paul, Trench, Trubner, 1924) was published in England by the International Library of Psychology, Philosophy, and Scientific Method, edited by C.K. Ogden. Burtt had tried, unsuccessfully, to have the volume published in the U.S. Since the book was a Ph.D. thesis and Columbia required 100 copies before granting Burtt his degree, he finally found a New York publisher to make the printing in the U. S. Harcourt, Brace & Company gave part of their issue a separate cover which reads: "*The Metaphysics of Sir Isaac Newton: An Essay on the Metaphysical Foundations of Modern Science*, by Edwin Arthur Burtt, AB, STM, Assistant Professor of Philosophy in the University of Chicago. Submitted in partial fulfillment of the requirements for the degree of Doctor of Philosophy, in the Faculty of Philosophy, Columbia University." The rest of the edition bore the title: *The Metaphysical Foundations of Modern Physical Science* (1925). The book was reissued in 1932 by Routledge & Kegan Paul in the second edition. Only the last chapter was changed, and it was completely rewritten with greater emphasis on the need for a new philosophy of mind. Since 1932 the book has been reprinted many times without further revision, and remains in print today. The Humanities Press picked it up in 1952, Anchor Books edition, 1954. First paper back edition, 1980. The sixth printing appeared in 1992. Some reprint editions have an error in the title, which reads, "*The Metaphysical Foundations of Modern Science.*" The original publisher is bringing out a new reprint edition in the near future. All citations in this book are from the 1992 edition, unless otherwise noted.

2 Burtt was reacting against the positivism of Ernst Mach, which had inspired the logical positivists of the Vienna Circle. Although "logical positivism' is the label which A.E. Blumberg and Herbert Feigl gave to the set of ideas associated with the Vienna Circle in 1931, the same name is used more broadly to include the analytical philosophies developed at Cambridge and Oxford. Burtt was immersed in teaching British empirical philosophy at the time he was writing *The Metaphysical Foundations.* Naturally he was discussing and debating the new realism of Whitehead, Russell and Moore. He must have been aware of the New Realism emerging in the United States, led in part by his thesis advisor, F.J.E. Woodbridge at Columbia. But primarily, it is the new British philosophy to which Burtt refers in his introduction. Therefore, it is

"logical empiricism" he is directly criticizing. "Logical empiricism" is the term of choice for Realists to describe their philosophy and its methods. Those less sympathetic call it "logical positivism". Here, in discussing the analytical philosophy from Burtt's point of view, the term "logical positivism" will be used. Otherwise, "logical empiricism" or "British empiricism" should be construed to mean the philosophy inspired by Bertrand Russell and A.N. Whitehead, married with Mach's ideas to become many species of analytical philosophy which dominated the American universities throughout most of the twentieth century.

[3]Daston, Lorraine. *History of Science in an Elegiac Mode, E.A. Burtt's Metaphysical Foundations of Modern Physical Science Revisited, ISIS* 82 (1991): 522-523 and 530-531.

[4] *Metaphysical Foundations,* 238.

[5] *Metaphysical Foundations,* 300.

[6] Thomas Kuhn, personal correspondence, 4 January 1994.

[7] Henry Guerlac, *Essays and Papers in the History of Modern Science,* (Baltimore: Johns Hopkins University Press, 1977), 63.

[8] Alexandre Koyré, *De la mystique a las science Cours, conferences et documents 1922-1962,* edited by Pietro Redondi, (Paris: Editions De L'Ecole Des Hautes Etudes en Sciences Sociales, 1986) Preface, XX-XXII.

[9] Thomas S. Kuhn, *The Structure of Scientific Revolutions,* second edition, *Preface,* (Chicago: University of Chicago Press, 1962, 1970), note 1, vi and T.S. Kuhn, *Alexandre Koyré & the History of Science, On an Intellectual Revolution, Encounter,* (Jan., 1970): 67-69.

[10] *Metaphysical Foundations,* 306-307.

11 Paul Hoyningen-Huene, *Reconstructing Scientific Revolutions, Thomas S. Kuhn's Philosophy of Science.* Translated by Alexander T. Levine with a Foreword by Thomas S. Kuhn. (Chicago and London: University of Chicago Press, 1993), 123.

12 Hoyningen-Huene, 130.

13 See H. Floris Cohen, *The Scientific Revolution* , 122-150, and especially, 130-131.

14 *Metaphysical Foundations,* 324-325.

Notes to Chapter 1

[1] Morris R. Cohen, *American Thought,* (Glencoe, Illinois: The Free Press, 1954), 74.

[2] Marcus G. Singer, *American Philosophy,* (Cambridge, London and New York: Cambridge University Press, 1985), 299.

[3] Ibid., 302.

[4] Some of these were Sidney Hook, Ernest Nagel, Herbert Schneider, Paul Weiss and Morton White.

[5] Morton White, *The Revolt Against Formalism in American Social Thought of the Twentieth Century, Journal of the History of Ideas,* Volume VIII, Number 2, (April,

1947): 131-152. White characterized the teachers of these men as "under the spell of history and culture".

Notes to Chapter 2

1 John Dewey, *Reconstruction in Philosophy,* (New York: Henry Holt and Co., 1920), 18.

2 J.H. Robinson, *The Mind in the Making,* 3.

3 John Dewey, *Reconstruction in Philosophy,* 2.

4 Burtt, *The Philosophy of Man as All-embracing Philosophy, The Philosophical Forum,* Volume II, number 2, (Winter, 1970-71), 162.

5 Ibid., 161.

6 Ibid., 170.

7 Ibid., 169.

8 J.H. Robinson, *The Mind in the Making,* (New York and London: Harper and Brothers, 1921), 108.

9 Harvey Wish, *Introduction* to J.H. Robinson, *The New History,* (New York: Macmillan Co., originally published 1912, 1965 edition), ix.

10 Ibid., xii.

11 Ibid., xv.

12 John H. Randall, *The Department of Philosophy* in *A History of the Faculty of Philosophy Columbia University,* (New York: Columbia University Press, 1957), 116.

13 Ibid., 130

14 Ibid., 127.

15 E.A. Burtt, *My Path to Philosophy, Philosophy East and West,* Volume 22, Number 4 (October 1972): 430.

16 Samuel Meyer, editor, *Types of Thinking including A Survey of Greek Philosophy by John Dewey,* (New York: Philosophical Library, 1984). Although Meyer claims this previously unpublished manuscript, documenting Dewey's history of philosophy and comparison of the various schools, is new, the book seems to be a popularized version of *Reconstruction in Philosophy* (1920), Dewey's most popular book. Richard J. Bernstein, writing for *The Encyclopedia of Philosophy* claims the work is based on the Japan/China lectures.

17 James Gutmann, *John Herman Randall, Jr.: A Memoir of His Career at Columbia, 1915-1967* in *Naturalism and Historical Understanding, Essays on the Philosophy of John Herman Randall,* Jr., John P. Anton, editor, (New York: State University of New York Press), 284.

18 Howard Thurman, *With Head and Heart,* (New York and London: Harcourt Brace Jovanovich, 1979), 44. I am indebted to Helen and Karl Schantz and Caroline and Frank Pineo of Ithaca,

New York for leading me to this source.

[19] John H. Randall, *How Philosophy Uses Its Past,* (New York: Columbia University Press, 1963), 18.

[20] Robinson resigned in 1919 to found the New School with Dewey and Beard, but Beard had resigned earlier, in 1917, to protest the firings of Professors J. M. Cattell and H.W. L Dana for their alleged support of pacifism. Joseph Freeman, *An American Testament,* (New York: Octagon Books,
1973): 104-109.

[21] J.H. Randall, Jr., Autobiographical Sketch preceding the essay, *Historical Naturalism* in *American Philosophy Today and Tomorrow,* Horace Kallan and Sidney Hook, editors, (New York: Lee Furman, Inc., 1935), 411.

[22] John Dewey, *Reconstruction in Philosophy,* (New York: Henry Holt and Company, 1920), 17.

[23] J.H. Randall, Jr., The Department of Philosophy in *A History of the Faculty of Philosophy Columbia University,* 129.

[24] John H. Randall, *The Department of Philosophy,* 127.

[25] Dewey, *Reconstruction in Philosophy,* 74-75.

[26] Ibid., 75.

[27] Ibid., 131.

[28] Ibid., 1.

[29] Ibid., 8.

[30] Ibid., 2.

[31] Ibid., 9.

[32] Ibid., 25.

[33] Ibid., 25.

[34] Ibid., Chapter 4.

[35] Ibid., 130-131.

[36] Ibid., 133.

[37] John Dewey, *How We Think,* (Boston, New York and Chicago: D.C. Heath, 1910), 6.

[38] Ibid., 9.

[39] Ibid., 12-13.

[40] Ibid., 13.

[41] Paul Kurtz, *American Philosophy in the Twentieth Century, A Source book from Pragmatism to Philosophical Analysis,* (New York and London: Macmillan Co., 1966 and 1968), 20.

[42] Dewey, *How We Think,* 22.

[43] *Metaphysical Foundations,* 229.

[44] E.A. Burtt, *The Contemporary Significance of Newton's Metaphysics* in *Isaac Newton 1642-1727, A Memorial Volume,* edited for the Mathematical Association by W.J. Greenstreet, (London: G. Bell and Sons, Limited, 1927), 140.

[45] *Metaphysical Foundations,* 324-325.

Chapter 3

[1] *Masterpieces of World Philosophy in Summary Form,* edited by Frank N. Magill and Ian P. McGreal, (New York: Harper & Brothers, 19961),1022.

[2] Ibid., 1022.

[3] Ibid., 1023.

[4]Sidney Hook, *Out of Step,* (New York: Harper and Row), 85-86.

[5] C.F. Delaney, *Mind and Nature,* (Notre Dame: U. of Notre Dame Press, 1969), 19.

[6] Randall, *The Department of Philosophy,* 119-120.

[7] Ibid., 123.

[8] Ibid., 119.

[9] F.J.E. Woodbridge, *The Purpose of History,* (New York: Columbia University Press, 1916), 27-28.

[10] Joseph Freeman, *An American Testament,* (New York: Octagon Books, 1973), chapter 1, "Saviors in Cap and Gown" and especially page 106 for a description of history at Columbia during the second decade of the century.

[11] Frederick F.J.E.. Woodbridge, *Nature and Mind: Selected Essays,* (New York: Russell and Russell, 1965), p. 283.

[12] Frederick J.W. Woodbridge, *An Essay on Nature,* (New York: Columbia University Press, 1940), p. 332.

[13] Magill and McGreal, 1022-1023, and Randall, *The Department of Philosophy,* 118.

[14]See Gary Hatfield, *Metaphysics and the New Science* in *Reappraisals of the Scientific Revolution,* ed. by David Linberg and Robert S. Westman. Cambridge/New York: Cambridge U. Press, 1990, pp. 93-166, and Lorraine Daston, *History of Science in an Elegiac Mode, E.A. Burtt's* Metaphysical Foundations of Modern Physical Science *Revisited. Isis* 82 (1991): 522-531.

[15] *Metaphysical Foundations,* 134.

[16] *Metaphysical Foundations,* 324.

[17] *Metaphysical Foundations,* 304

[18] *Burtt, Types of Religious Philosophy,* (New York: Harper and Brothers, 1939 and revised 1951) revised edition, 288.

[19] Ibid., 24.

[20] See Lorraine Daston.

[21] E.A. Burtt, *My Path to Philosophy,* 431.

[22] J.H. Randall, Jr., *The Department of Philosophy,* 134.

[23] Sidney Hook, 60.

[24] Ibid., 56.

25 Ibid., 58.

26 David Hollinger, IEthnic Diversity, Comopolitanism and the Emergence of the American Liberal Intelligentsia. *American Quarterly,* XXVII (1975): 135.

27 Ibid.

28 Sidney Hook, 60.

29 David Hollinger, *Morris R. Cohen and the Scientific Ideal.* (Cambridge: MIT Press), 234.

30 Ibid. 235.

31 Ibid., 55-56.

32 Ibid., 55.

33 Hollinger quotes these lines from many diverse sources. See his footnotes 38 and 43. See pages 57, 64, 65.

34 Quoted from Russell's *A Free Man's Worship* in *The Metaphysical Foundations,* 23.

35 *The Metaphysical Foundations,* 229.

36 C.F. Delaney, *Morris R. Cohen, A Preface to Logic,* (Meridian Books, Cleveland: The World Publishing Co., 1956) 19.

37 Quoted by Delaney, 20.

38 Burtt, *Discussion, Review of M.R. Cohen, Reason and Nature* in *The Philosophical Review,* Volume 41 (1932): 316-317.

39 Ibid., 617.

40 Ibid., 618-619.

41 *Metaphysical Foundations,* 330.

42 Quoted from Morris Cohen, *Reason and Nature,* 165 by Marcus Singer in *American Philosophy,* (Cambridge and London: Cambridge University Press, 1985), 305.

43 Harcourt, Brace and Company is now located in San Diego, California. A telephone call confirmed that the original edition of the book is in the archives, number 3128. The frontpiece reads: "Routledge-Kegan Paul Trench and Company, London and Harcourt, Brace and Company, New York."

Chapter 4

1 H. Floris Cohen has given the best account and critique of E.A. Burtt and *The Metaphysical Foundations* in print today in his *The Scientific Revolution, An Historiographical Inquiry,* (Chicago: University of Chicago Press, 1994). Cohen is the first to acknowledge the significance of Burtt's "radical historical critique," where others have missed it. Cohen is liberal with his credit to Burtt as "the first to adopt a 'relative discontinuity' position" to explain how scientific knowledge is acquired (96, 149) and as "at the head of an array of historians to the Scientific Revolution who, in the sixties, began to look upon this signal event in European intellectual history as having brought, besides impressive gains in our knowledge, some no less consequential losses" (96-97). Although Cohen acknowledges that it is Alexandre Koyré who is at the center of the

historiographical "treasure" that the Scientific Revolution has become, he makes it clear that Koyré's priority in this treasure is a debt he owed to Burtt (101). Most important of all, Cohen did not miss the fact that Burtt's work has been a powerful influence in "the debate over the rationality of early modern science." He rightly points out the wide-ranging impact of Burtt's work on T.S. Kuhn and Frances Yates (111, 179-182) and sums up this way: "Lonely Burtt's account is still very relevant to the course of present-day discussions of the Scientific Revolution. And what is even more: It is from Burtt's book that Koyré's concept of the Scientific Revolution probably found its origin" (89).

2 Burtt, *Two Basic Issues In the Problem of Meaning and Truth* in *Essays in Honor of John Dewey on the occasion of his seventieth birthday*, October 20, 1929, (New York: Henry Holt and Company, 1929), 69.

3 R.G. Collingwood, *Essay on Metaphysics*, (Oxford, England: The Clarendon Press, 1940).

4 The term Burtt used here was used both by John Dewey in his *Reconstruction in Philosophy* and by the Phenomenologists who borrowed it from Francis Bacon.

5 *Metaphysical Foundations*, 229.

6 Ibid., 229.

7 Ibid., 229.

8 Ibid., 229.

9 Ibid., 228.

10 Ibid., 228.

11 Ibid., 228.

12 Ibid., 228-229.

13 Ibid., 227-228.

14 *Metaphysical Foundations*, 227.

15 Elaine Daston in *Isis*, 1994.

16 Burtt's historical approach earned him the misnomer, Hegelian, but at no time in his long career did Burtt ever believe that there was an Absolute and True Reality or Being qua Being which man could know. His preference for the genetic method led him to a Collingwood brand of historicism and his interest in religion invited the inaccurate conclusion that he was a "true believer." He was agnostic about the nature and even the existence of the traditional God of Christianity until the last years of his life.

17 Burtt, *The English Philosophers from Bacon to Mill*, (New York: The Modern Library, 1939), xi-xii.

18 *The Metaphysical Foundations*, 208.

19 Quoted from Bertrand Russell, *A Free Man's Worship* in *Mysticism and Logic* (New York, 1918) *Metaphysical Foundations*, 23.

20 Russell, *A Free Man's Worship* in *Philosophical Essays* , (New York, Bombay, and Calcutta: Longmans, Green, and Co., 1910), 61.

21 Ibid.

[22] John H. Randall, *The Making of the Modern Mind,* (Boston, New York, etc.: Houghton Mifflin Company, 1926), 637.

[23] John H. Randall, *The Making of the Modern Mind,* 638.

[24] Even further background can be gained on the impact of science during the last of the nineteenth and the early part of the twentieth centuries in Randall's *Philosophy After Darwin Chapters for The Career of Philosophy* Volume III and Other Essays, edited by Beth Singer, (New York: Columbia University Press, 1977). See the first three chapters, especially, Chapter 1, *The Conflict of the Religious Tradition With Science* and page 29 where Randall discusses Bertrand Russell's *A Free Man's Worship* and what Russell's essay meant to the young radicals around Columbia University in the 1920s.

[25] Bertrand Russell, *Two Dogmas of Naturalism,* a review of *The Metaphysical Foundations of Modern Physical Science* by E.A. Burtt in *Dial* 79 (September, 1925): 255-258.

[26] Russell, *History of Western Philosophy and it Connection with Political and Social Circumstances from the Earliest Times to the Present Day,* (London: George Allen and Unwin Ltd., first published in 1946). New edition (reset), 1961 is cited here, pages 514-515.

[27] Ibid., 210.

[28] Ibid., 212.

[29] Ibid., 214.

[30] Ibid., 212.

[31] Ibid., 230.

[32] Ibid., 218. Quoted from Newton's *Principles,* II, 314. Burtt's further notation is: Cf. also *Opticks,* 380.

[33] Quoted from Newton's *Principles,* II, 160 in Ibid., 218-219.

[34] *Metaphysical Foundations,* 229.

[35] Ibid., 229.

[36] See James E. Force, *Newton's Sleeping Argument and the Newtonian Synthesis of Science and Religion* in *Standing on the Shoulders of Giants, A longer View of Newton and Halley. Essays Commemorating the 1985-1986 Return of Comet Halley,* edited by Norman J.W. Thrower, (Berkeley: University of California Press, 1990), page 125 and Force, *Hume's Interest in Newton and Science* in *Essays on the Context, Nature, and Influence of Isaac Newton's Theology,* edited by James E. Force and Richard H. Poplin, (Dordrecth, Boston and London: Kluwer Academic Publishers, 1990), 181-206; see especially page 199, footnote 30. Force explains how his interpretation of Newton relies on Burtt. Force has taken Burtt's reading of Newton and Hume, which involves Burtt's own undeclared use of Hume's brand of radical skepticism turned back on Hume to give what Force wants to see as Hume's reading of Newton. It is Burtt's reading of Hume and Burtt's reading of Newton which Force takes as Hume's critique of Newton.

[37] Force, *Newton's Sleeping Argument and the Newtonian Synthesis of Science and Religion,* 125.

[38] David Hume, *Concerning Natural Religion* Part II in *The English Philosophers from Bacon to Mill,* edited and with an Introduction by E.A. Burtt, (New York: Modern Library, Random House, 1939), 699-708 and Part XII, 753-755.

[39] *Metaphysical Foundations,* 300.

[40] H. Floris Cohen, 92.

Chapter 5

[1] Darnell Rucker, *The Chicago Pragmatists,* (Minneapolis: University of Minneapolis Press, 1969), 6. Rucker says, "Albion Small, for instance, left Colby to found the first graduate Department of Sociology in the world. Dewey came two years later despite what he considered an inadequate salary because Chicago offered to let him work in psychology and education, as well as philosophy, an opportunity that he could not expect at Michigan.

[2] T.V. Smith, *A Non-Existent Man,* (Austin, Texas: University of Texas Press, 1962), 53.

[3] Darnell Rucker, 24.

[4] T.V. Smith, 51.

[5] Darnell Rucker, 26.

[6] T.V. Smith, 45-46.

[7] Ibid., 49.

[8] Gary Cook, *George Herbert Mead, The Making of a Social Pragmatist* , (Urbana and Chicago, Illinois: University of Chicago Press, 1993), 221-213.

[9] John Herman Randall, Jr., *The Department of Philosophy* in *A History of The Faculty of Philosophy, Columbia University,* (New York: Columbia University Press, 1957), 130.

[10] William H. McNeill, *Hutchins' University,* (Chicago: University of Chicago Press, 1991), 16-17.

[11] Ibid., vii and 20.

[12] Cook, 184.

[13] Quoted from Tufts' memorandum in Cook, 185.

[14] Quoted by Cook, 186.

[15] Burtt, *A Statement From The Department of Philosophy,* 1.

[16] M.J. Adler, *Philosopher at Large: An Intellectual Autobiography.* (New York: Macmillan, 1977), 130-133.

[17] Burtt, *A Statement From The Department of Philosophy,* 3.

[18] *A Statement From the Department of Philosophy* filed under 12/3/1931 as *A Statement of Professors, Mead, Burtt and Murphy,* File: The Faculty of the Division of the Humanities: December, 1930-October, 1940. Kept in the Secretary of the Faculties Office at The University of Chicago.

[19]Cook, 192-3.

[20]G.J. Laing, *The Situation In The Department of Philosophy*, File: The Faculty of the Division of the Humanities: December, 1930-October, 1940. Kept in the Secretary of the Faculties Office at The University of Chicago.

[21]Marcus G. Singer, *Memoir* in Arthur E. Murphy, *Reason, Reality, and Speculative Philosophy,* Madison, WI: University of Wisconsin Press, 1996), xxi.

[22]This was due in part to Murphy and also to Norman Malcolm's personal friendship with Wittgenstein. Reading Marcus Singer in ibid. is a good place to catch an inkling of the pride and passion for Wittgenstein which caught the Cornell philosophers (except Burtt) after Murphy. Even Gregory Vlastos, a scholar of Greek Philosophy, caught the bandwagon. In 1949 Wittgenstein went to Ithaca and stayed with Norman Malcolm, whom Murphy had brought in from Princeton. Singer recalls Wittgenstein's presence and influence within the Cornell Philosophy Department. See Ibid., xxvii-xxviii.

[23] J. Copeland to D. Villemaire, December 6, 1995.

[24]John Nelson to Diane Villemaire, 12/7/95.

[25]Bernard Wand to Diane Villemaire, December 19, 1995.

[26]McNeill, 6.

[27]Smith, 48.

[28]Hollinger, *The Problem of Pragmatism in American History. Journal of American History.* 67 Number 1 (June, 1980): 89.

Chapter 6

[1]Edwin H. Wilson, *Genesis of A Humanist Manifesto,* (Amherst, NY.: The Humanist Press, 1995) can be downloaded from http://www.infidels.org/library/modern/ edwin_wilson/manifesto. Wilson was one of the original group and has written his memoir of the era. *The New Humanist* ceased publication in 1936 due to lack of funds. Wilson continued to produce a version of it, personally funded, *The Humanist Bulletin.* In 1941 *The Humanist* magazine began its series, still in print today.

[2] Edwin Wilson, *The Genesis of A Humanist Manifesto,* Chapter 1, page 1.

[3]Ibid., Chapter 5, page 5.

[4] Burtt, *Religion in an Age of Science*, (New York: Henry Holt and Company, 1929), front piece.

[5]Burtt, "Two Basic Issues in the Problem of Meaning and of Truth" in *Essays in Honor of John Dewey on the occasion of his seventieth birthday*, October 20, 1929, (New York: Henry Holt and Company, 1929), 73. Burtt made the argument in *Religion in an Age of Science,* and then argued similarly in a shorter version for the essay quoted in this text.

[6] *Religion In an Age of Science*, 152.

[7]Ibid. Both the manifesto and the article by Sellars appeared in the May/June issue of The New Humanist (VI: 3):1933.

[8]Reprinted in Wilson, Chapter 6, pages 1 and 2.

[9] This whole line was changed in the final draft.

[10] Adopted

[11] Rejected

[12] Adopted

[13] Rewritten for the final draft to read: "Holding an organic view of life, humanists find that the traditional dualism of mind and body must be rejected."

[14] Rewritten to incorporate Burtt's sense of transcendent realities, but weighted toward the scientific, away from the religious. Adopted for the final draft: "Humanism asserts that the nature of the universe depicted by modern science makes unacceptable any supernatural or cosmic guarantees of human values. Obviously humanism does not deny the possibility of realities as yet undiscovered, but it does insist that the way to determine the existence and value of any and all realities is by means of intelligent inquiry and by the assessment of their relation to human needs. Religion must formulate its hopes and plans in the light of scientific spirit and method."

[15] Rejected. This suggestion was key to Burtt's own thinking at the time and to the later program he wanted to work out, a scientific metaphysics of categories. Burtt intended to describe the categories of mind adopted from the scientific world view and then to apply the same descriptive method to the categories of religion such as God, salvation, soul, and later Burtt would add sin to the list. The point would be to determine "the meaning and value" of "realities transcending human [sense] experience", but still very much a part of the human experience.

[16] Added to the final draft as: "Certainly religious institutions, their ritualistic forms, ecclesiastical methods, and communal activities must be reconstituted as rapidly as experience allows, in order to function effectively in the modern world." Here Burtt's intention that philosophers should guide the progress of civilization through the scientific age toward an era where the spirit of man might regain a place in the world-view can be seen clearly.

[17] Added to the final draft as: "The goal of humanism is a free and universal society in which people voluntarily and intelligently cooperate for the common good."

[18] Ibid.

[19] Ibid. page 4.

[20] Ibid., chapter 9, page 2.

[21] Ibid., chapter 9, pages 3-4.

[22] Burtt, *Mind, Matter and Evolution* in *Mind and Behavior,* Baker Brownell, Editor, *Man and His World Series,* Volume III, (New York: D. Van Nostrand Co., Inc., 1929). 151..

[23] Ibid., 155.

[24] Ibid. 160-1.

[25] Ibid., 164

[26] Wilson, Chapter 9, page 4.

[27] Wilson, Preface, page 3.

[28] Corliss Lamont, *Voice In the Wilderness,* (Buffalo, NY.: Prometheus Books, 1974), 3-9.

Chapter 7

[1] Burtt, *Review of* *Matter and Gravity in Newton's Physical Philosophy. A Study in the Natural Philosophy of Newton's Time,* *The Journal of Philosophy* Volume 24, (1927): 670.

[2] Burtt, *Real vs. Abstract Evolution, Proceedings of the Sixth International Congress of Philosophy.* Harvard University, Cambridge, Massachusetts, September 13, 14, 15, 16, 17, 1926. Edgar Sheffield Brightman, editor, (New York and London: Longmans, Green, 1927), 170-171.

[3] P. Duhem, *The Aim and Structure of Physical Reality,* (New York: Athenaeum Books, 1977), 16-18. First edition, 1906.

[4] Burtt, *Real vs. Abstract Evolution,* 176.

[5] Ibid., 177.

[6] Ibid., 176.

[7] Burtt's treatment of time in 1926 is similar to R.G. Collingwood's idea of history and his argument that the past is not real. Only mind thinking the events and ideas of the past in the present and working out solutions to problems raised by the events and ideas of the past is real. Collingwood's sense of the real here is Burtt's abstract time, while Collingwood's idea of historical events as not real, corresponds to what Burtt calls real time. With that in mind, Burtt's conclusion can be compared to Collingwood's philosophy and to historicism in general. Collingwood's *Idea of History* was published posthumously in 1946. It was based on lectures Collingwood delivered at Oxford from 1926 through the 1930s.

[8] *Real vs. Abstract Evolution,* 177.

[9] Burtt, *The Generic Definition of Philosophic Terms, The Philosophical Review,* Volume LXII, No. 1, (January, 1953): 41. A paper read before the Philosophical Club of

Cornell University, November 29, 1951, and before the Philosophical Forum of Princeton University on December 4, 1951.

[10]*Metaphysical Foundations*, 307.

[11]Ibid.

[12] Burtt, *The Contemporary Significance of Newton's Metaphysics* in *Isaac Newton 1642-1927, A Memorial Volume,* edited for the Mathematical Association, (London: G. Bell and Sons limited, 1927), 140.

[13] Ibid., 137.

[14] Ibid., 138.

[15] Ibid., 138-139.

[16] Ibid., 140.

[17] Ibid., 139.

[18] Ibid.

[19] *Real versus Abstract Evolution,* 173.

[20] Ibid., 176.

[21] *Genesis of Hypothesis,* an unpublished manuscript, Rare and Manuscript Collections, Carl A. Kroch Library, Cornell University, Ithaca, New York, 1935, filed in Box 2 #14/21/2593, pages 122-123. See also *The Philosophy of Man as All-embracing Philosophy,* The Philosophical Forum, Volume II, number 2, (Winter 1970-71), 161 for Burtt's opinion on Kant's mistakes in formulating his metaphysics or "critical philosophy." Burtt argues with Kant's assumption that objective knowledge requires a universal mind or transcendental consciousness ending in a final solution to the problems of knowledge. He says later that "the structure of mind is actually relative and transitory" (162). This is so because mind is seen as evolving to meet the developing needs of man as his consciousness expands toward a realization of the spirit. (See *The Human Journey* (1981).

[22]Lawrence Cahoone, Editor, *From Modernism to Postmodernism,* (Cambridge, Massachusetts: Blackwell Publishers Inc., 1996), 15.

[23]This is a term Burtt used for a brief period around 1932. He might have been affected by "Operationalism" in science. Percy William Bridgman, the physicist who won the Nobel Prize in 1946, had published *The Logic of Modern Physics* in 1927. Burtt was keenly interested in such matters in 1926 through 1935 when he used the term. Bridgman's operationalism is supposed to be predicated not on philosophy at all, but on the methods employed by scientists themselves. Burtt was an advocate that philosophy should adopt the methods of science. Bridgman's operationalism has been criticized as psychological. It aims at clearing away all presuppositional material from scientific terms and articulating in full detail the criterion for meaningfulness of each term before accepting it into use. There are some similarities between Burtt's method for cooperation among philosophers and Bridgman's operationalism. Bridgman did think that his ideas had application in a wider setting, solving social problems, for example. It is not known if Burtt was familiar with Bridgman, but it seems likely.

[24]Burtt, *Present Day Tendencies in Ethical Theory, International Journal of Ethics,* Volume 31 (July, 1921): 432-438.

[25]Ibid., 438.

[26]Burtt, *Religion In An Age of Science,* 1929.

[27]Burtt, *Review of Matter and Gravity in Newton's Physical Philosophy. A Study in the Natural Philosophy of Newton's Time, The Journal of Philosophy* Volume 24, (1927): 670.

[28]Burtt, *Two Basic Issues In the Problem and Meaning of Truth* in *Essays in Honor of John Dewey On the Occasion of His Seventieth Birthday, October 20, 1929,* (New York: Henry Hold and Company, 1929), 74-75.

[29]Burtt, *Religion In An Age of Science,* (1929), 123.

[30]Ibid., 125.

[31]*Two Basic Issues In the Problem and Meaning of Truth,* 75-76.

[32]Ibid., 76,

[33]Ibid., 78-79.

[34]See Burtt's Review of M.R. Cohen, *Reason and Nature, Philosophical Review* 41 (November, 1932), 619. Burtt says that Pragmatism is "a confused movement on its metaphysical side," referring specifically to the pragmatists' unwillingness to examine the conditions under which the same phenomenon might be experienced differently by different individuals.

[35]Burtt, *The Contemporary Significance of Newton's Metaphysics,* 140.

[36]See below and my master's thesis, *The metaphysics in the Twentieth Century the Philosophies of R.G. Collingwood and E.A. Burtt,* University of Vermont, 1994.

[37]Burtt, *Relativity and Scientific Method in Philosophy* published in *College of the Pacific Publication in Philosophy, Volume 1. Lectures Delivered Under the Auspices of the Pacific Philosophy Club in Anderson Hall During 1931-1932,* edited by Paul Arthur Schilpp, (Stockton, California: College of the Pacific, 1932), 99.

[38]Ibid. , 103-104.

[39]Burtt, *Two Basic Issues In the Problem and Meaning of Truth,* 75-76.

[40]Burtt, *British Philosophers From Bacon to Mill,* xi.

[41]Burtt, *The Contemporary Significance of Newton's Metaphysics,* 26.

[42]Burtt, *Review of Reason and Nature,* 617-618.

[43]Ibid., 618

[44]Ibid., 619

[45]Ibid.

[46]Ibid., 620.

[47]See Burtt, *The Human Journey.* Stephanos Nirmalendu Ghosh Lectures on Comparative Religion delivered in the Ramakrishna Mission Institute of Culture for the University of Calcutta in January, 1968. Calcutta, India: University of Calcutta Press, 1981.

[48]*The Human Journey*, 29-32.

[49]The best presentation of these ideas is to be found in *The Philosophy of Man, an All-embracing Philosophy* written in 1969, published in 1970-71.

[50] *Metaphysical Foundations*, 134.

[51] Ibid., 304.

Chapter 8

[1]All references in this chapter to the manuscript, *The Genesis of Hypothesis* are noted as (GH) with the page number directly in the text.

[2]This idea recalls Burtt's argument in his review of Cohen's *Reason and Nature*, which had been an obvious response to Dewey's *Experience and Nature*.

[3] Handwritten at the bottom of the page is Burtt's note: "Thus our conclusion = laws of non thinking nature depend in ways above on thinking nature." Unfortunately Burtt never did pursue the consequences of this equation, but what he seems to have been playing with are some notions more recently put forward by John Barrow and Roger Penrose, among others, which have been called strong and weak "Anthropic Principle". See Chapter Eight in John Barrow, *Theories of Everything* (Oxford: The Clarendon Press, 1991). or Stephen Weinberg, *Dreams of a Final Theory, The Scientist's Search for the Ultimate Laws of Nature*, (New York: Vintage Books, 1992,)

[4] This compares to Burtt's lecture at the Philosophy Club meeting at the College of the Pacific in 1931.

[5]Burtt, *Relativity and Scientific Method in Philosophy*, 89-90.

[6]Ibid., 90.

[7] Burtt, *Metaphysics of Empirical Theology* , manuscript of seven numbered pages and one additional page inserted. Rare and Manuscript Collection, Olin Library, Cornell University, Ithaca, New York.

[8] *Metaphysics of Empirical Theology* , Point 9, page 4.

[9]Ibid., Point 11, page 5.

[10] Burtt, *Value and Existence, Journal of Philosophy*, Volume 44, Number 7 (March, 1947).

[11] Ibid., 177.

[12]David Hollinger, *T.S. Kuhn's Theory of Science and Its Implications for History, The American Historical Review*, 78 (1973): 390-391.

Chapter 9

[1] Burtt used the 1803 Motte translation of Newton's *Principia*. He read the *Opticks* and the *Arithmetic*, plus the *Optical Lectures* and the *Opera Omnia* edited by Horsley. He examined the Catalogue of the Portsmouth Collection of Newton's books and papers and Gray's 1907 bibliography of the Works of Sir Isaac and had looked at all the published correspondence available at the time in Richard Bentley, *Correspondence*, edited by Wordsworth (1842) and J. Edleston's *Correspondence of Sir I. Newton and Prof. Cotes* (1850).

[2] *Metaphysical Foundations, Introduction.*

[3] Ibid., 203-4.

[4] Ibid., 226.

[5] *Metaphysical Foundations*, 256.

[6] Ibid., 256-257.

[7] Ibid., 258.

[8] Ibid, 261.

[9] Ibid, 282.

[10] Ibid., 288.

[11] Ibid., 291.

[12] Ibid., 262.

[13] Ibid., 238 and 300.

[14] Reprinted in Appendix A to *The Conway Letters*, edited by Marjorie Hope Nicholson, (Oxford: Clarendon Press, 1992), 478-479.

[15] R. Hooykaas, *Religion and the Rise of Modern Science*, (Edinburg and London: Scottish Academic Press, 1972), 34. The passage is quoted, noting Oresme, without a citation.

[16] See Lorraine Daston, *History of Science in an Elegiac Mode*, Isis, 82, (1991): 522-531 and Richard S. Westfall, *Science and Religion in Seventeenth-Century England*, (New Haven: Yale University Press, 1958), 12.

[17] *Metaphysical Foundations*, 34-35.

[18] See Burtt's article, *The Generic Definition of Philosophic Terms*, 41.

[19] Burtt, *Two Basic Issues In the Problem of Meaning and of Truth*, 69.

[20] Sir David Brewster, *Memoirs of the Life, Writings and Discoveries of Sir Isaac Newton* (Edinburgh, 1855), 270.

[21] David Brewster, 272.

[22] Ibid., 283.

[23] Ibid., 250-251

[24] Ibid., 262.

[25] Ibid., 248-249.

[26] Ibid., 277.

[27] Ibid., 284.

[28] Ibid., 285.

[29] Ibid., 286.

[30] Ibid., 268

[31] Ibid., 286.

[32] *Metapysical Foundations*, 284.

[33] Ibid.

[34] Richard Westfall, *Science and Religion in Seventeenth-Century England*, (New Haven: Yale University Press), 193-220.

[35] *Metaphysical Foundations*, 283-284.

[36] Ibid, 300-301.

[37] An exception is Richard Westfall's 1958 *Science and Religion in Seventeenth-Century England*. Westfall saw Burtt then as a Christian apologist.

[38] Harry Elmer Barnes, *The Historian and the History of Science*, *The Scientific Monthly*, Volume 11, (1919): 113.

[39] Ibid.

[40] Barnes, *Does Science Require a New Conception of God?*, *Current History*, XXIX, number 6, (March, 1929): 883-898.

[41] John H. Randall, Jr., *Effects of Science and Industry on Religion*, *Current History*, XXX, Number 3, (June, 1929): 355-366.

[42] See Richard Westfall's Bibliographic Essay in *Science and Religion in Seventeenth-Century England* (1958) for a summary of historiography which treats the conflict from the middle of the nineteenth century onward.

[43] Cohen, 82-83.

[44] *Sir Isaac Newton, 1727-1927, A Bicentenary Evaluation of His Work*, (Baltimore: The Williams and Wilkins Co., 1928), viii-ix.

[45] Harry Elmer Barnes, "The Historian and the History of Science" *Scientific Monthly* Volume 11 (1919): 112.

[46] Arnold, Thackray. *History of Science* in *A Guide to The Culture of Science, Technology and Medicine*, Paul T. Durbin, editor, (New York: The Free Press, 1984), 13-14.

[47] Ibid.

[48] Westfall, 223.

[49] Ibid, 12.

[50] This is a trend in the history of science which remarkably was out of step with the mathematicians themselves and later out of step with a large group of philosophers of science. See *Mathematics and the Loss of Certainty* in John Gillott and Manjit Kumar, *Science and the Retreat From Reason*, (New York: Monthly Review Press, 1997), 209-213. The work of the logician Kurt Godel in 1931 is widely taken to undermine the possibility that truths about mathematics are truths about nature. "After 1931, most mathematicians decided that Godel's result put an end to the great attempts to put [mathematical science theories] on a rigorous axiomatic footing--attempts led by

Russell, Whitehead, and David Hilbert. Today this seems indisputable...." (Gillott and Kumar, 211).

[51]Strong's dissertation was published as *Procedures and Metaphysics, A Study in the Philosophy of Mathematical-Physical Science in the Sixteenth and Seventeenth Centuries,* (Berkeley, California: University of California Press, 1936). Burtt's 1932 revised *Metaphysical Foundations* is used as a point of departure for Strong, who builds his thesis around the Mathematical realism of Galileo and Kepler, without much attention to Newton. Strong wanted to separate the mathematical methods of science and the instrumental truth or science knowledge derived from them from any metaphysical underpinning which might have an historical basis. His argument rests on the opinion that "metaphysical assumptions" do not need to become tangled up with "methodological considerations" in science to justify science knowledge which is real in its own right....The scientist does not have to know what the universal conditions of knowing are in order to solve problems arising within his field of inquiry and to verify and demonstrate propositions within that field." (Ibid., 7). "The 'mathematical interpretation' of nature in a systematic philosophy is not mathematical procedure in physical investigation....The methods and subject matter of science...are the autonomous or self-justifying knowledge constituting the science...Its methods and results are real property with a clear title" which can not be made dependent upon a mathematical interpretation of nature as metaphysics. (Ibid., 8) See also *Newton's Mathematical Way, Journal of the History of Ideas,* XII, (January, 1951): 90-110. Here Strong argues against Burtt and John Randall, Jr., who "have argued that there is an unreconciled conflict in Newton's thought between his mathematical rationalism...and his empiricism." Strong argues that Newton was both a mathematical realist and an empiricist, who reconciled the two with his experimental method. In *Newton and God, Journal of the History of Ideas,* XIII, number 2 (1952): 147-167) Strong argues against Burtt's contention that "for Newton, 'the realm of science was dependent on the God of religion." Strong contends that there was no theological foundation to Newton's science. *Newtonian Explications of Natural Philosophy, Journal of the History of Ideas,* XVIII, number 1, (January, 1957): 49-83 makes no mention of Burtt. In Strong's *Hypotheses Non Fingo* published in *Men and Moments In the History of Science,* Herbert M. Evans, editor, (New York: Greenwood Press, 1959), 162-176 Newton's rules of reasoning, one through three, are offered as good argument for Newton's statement in the title. But Strong leaves out rule four, which is the one Burtt found qualified rule three. Strong has completely ignored Burtt's argument here, while taking up his subject and his proof. By abridging Burtt's proof, Strong escaped having to deal with Newton's uncertainty about the laws of the universe. Burtt is not mentioned in this article. But then by 1970 something interesting happens. Strong comes back to Burtt.

Strong's 1970 *Barrow and Newton, Journal of the History of Philosophy,* VIII, (April, 1970): 155-172 begins with a reference to Burtt and *The Metaphysical Foundations* in the first sentence. In 1962 A. Rupert and Marie Boas Hall had made previously unpublished Newton papers available indicating the indebtedness of Newton to Barrow. This information was not available to Burtt. Strong's paper confirms Burtt's contention that Newton's concept of space as some kind of extension of God derived from Barrow.

[52]Strong, it should be noted, was a philosopher who used the genetic method to illuminate his philosophical arguments, typical of the pragmatic thinkers educated at Columbia. Strong's work makes a good companion to *The Metaphysical Foundations* since both grew out of the same tradition. Both works are historically based. Strong argues for mathematical realism, representing the more formally minded pragmatists deriving from Peirce; Burtt argues against, representing the naturalist/evolutionist minded camp, deriving from John Dewey.

[53]Louis Trenchard More, *Isaac Newton. A Biography* (London: Constable and Co, 1934 and New York: Charles Scribner; New York: Dover, 1962).

[54]L.T. More, *Isaac Newton* (New York, 1962), 645.

[55]John Maynard Keynes, *Newton the Man* in *The Collected Writings of John Maynard Keynes Vol. X Essays in Biography,* (New York: Macmillan St. Martins Press for the Royal Economic Society, 1972, first published 1933), 363-364.

[56]Yates' work is *Giordano Bruno and the Hermetic Tradition* (1964) and an article, *The Hermetic Tradition in Renaissance Science* (1967) and Kuhn's work has already been mentioned, 1962.

[57] Cohen, 179 citing A.R. Hall in "Magic, Metaphysics, and Mysticism in the Scientific Revolution", published in *Reason, Experiment and Mysticism in the Scientific Revolution,* 277.

[58]Cohen, 180 citing P. Rossi in "Hermeticism, Rationality and the Scientific Revolution", published in *Reason, Experiment and Mysticism in the Scientific Revolution,* 272.

[59]Frances Yates, *Giordano Bruno and the Hermetic Tradition,* (Chicago: University of Chicago Press, 1964): 454-455.

[60]Cohen, 88-89.

[61]A. Rupert Hall, *The Rise and Fall of Science,* A review of H. Floris Cohen's *The Scientific Revolution, Nature* (November 3, 1994) My copy of this review is a photocopy of the review sent to Cohen by *Nature* editors.

[62]A.R. Hall, *The Revolution in Science 1500-1750,* (1983). In the introduction Hall claims himself to be unashamedly positivist or whigish. "I unashamedly follow a positivist or even whiggish line, for it is impossible to write in the same sentence of the victors' and the losers' view of a battle. I do not believe that Copernicus is an important figure in history because he once names Hermes, that Kepler's astrological tract is his most important work, that Newton's name is immortal because he read alchemical authors..." (page 2). Clearly, Hall holds a black and white, progressive view of science and believes the reality of the world is mathematical. He distinguishes between Platonic and Pythagorean mathematics, relying one hundred percent on Koyré's essay, "Newton, Galileo and Plato" without citing it.

[63]George Sarton, *The History of Science and the New Humanism,* (Cambridge, Massachusetts: Harvard University Press, 1937), 111.

[64] Ibid, 110.

[65] Ibid, 110.

[66] Cohen, 80.

[67]See Lorraine Daston, 1992.

[68]*Metaphysical Foundations*, 226.

[69]Hall (1983), 355.

[70]See B.J. Dobbs, *Hunting the Green Lyon*, (Cambridge University Press, 1975) and A. R. Hall's *Isaac Newton, Adventurer In Thought*, (London: Butler and Tanner Ltd., 1992) and *Henry More Magic Religion and Experiment*, (London: Butler and Tanner Ltd., 1990). Both authors acknowledge the facts while assigning them different levels of importance. Also see Scott Mandelbrote, *A duty of the greatest moment* , *British Journal for the History of Science*, 26 (1993): 281-302.

[71]Burtt offered many details of Newton's relationship to Barrow on pages 250-251 of *The Metaphysical Foundations*. However, all the associations with Henry More are only implied through careful study of the ideas of both More and Newton and detailed indications of where they overlap.

[72] *Metaphysical Foundations*, 238.

[73]Ibid 139.

[74]Ibid, 238.

[75]Ibid 299. For later historians agreeing with Burtt's view that Newton was indeed a transition figure and not a fully modern one, see Richard S. Westfall, *Science and Religion in Seventeenth-Century England*, (New Haven: Yale Press, 1958); R. Hooykaas, *Religion and the Rise of Modern Science*, (Edinburgh and London: Scottish Academic Press, 1972) and B.J. Dobbs, *Hunting the Green Lyon*.

[76]Hall (1983), 287, quoting Burtt in *The Metaphysical Foundations*.

[77]Ibid., 288.

[78]Ibid.

[79]Dobbs, *Newton as Final Cause and First Mover*, *Isis*, 85 (1994): 640-41.

[80]Hall (1990), 202.

[81]Ibid., 203.

[82]Ibid. 202-203

[83]Ibid., 203.

[84] See above, Chapter 1. Quoted from Burtt's paper, *Empirical Theology*.

[85]John K. Wood, *The Nature of Conflicts Between Science and Religion*, (Logan, Utah: Utah State University Press Monograph series, Volume IX, number 2, March, 1962), 12.

[86] Ibid., 12-13

[87] A copy of *The Philosophical Forum* in which an early article by Holton, *Johannes Kepler: A Case Study on the Interaction of Science, Metaphysics and Philosophy*, appears was found among Burtt's papers. The Journal is published by the Philosophical Club at Boston University. Peter Bertocci is listed as a member of the Advisory Committee.

[88] Gerald Holton, *Thematic Origins of Scientific Thought, Kepler to Einstein*, (Cambridge, Massachusetts: Harvard Press, 1988), 12-13.

[89]Ibid., 17.

---

[90]See B.J.T. Dobbs, *Newton as Final Cause and First Mover*, Isis, 85 (1994): 633-643; James E. Force, *Newton's Sleeping Argument and the Newtonian Synthesis of Science and Religion* in *Standing On the Shoulders of Giants, A Longer View of Newton and Halley*, edited by Norman Thrower, (Berkeley, California: University of California Press, 1990); and for more moderate views: David Kubrin, *Newton and the Cyclical Cosmos: Providence and the Mechanical Philosophy, Journal of the History of Ideas* 28 (1967): 325-46 and Loup Verlet, *The Newtonian Revolution, History of Science* volume 34, number 3 (September, 1996): 303-344. A new biography has just been written by Michael White of England, *Isaac Newton, the Last Sorcerer*, (Reading, MA: Helix Books and Addison-Wesley, January, 1998). White has been extensively interviewed in the United States as if his Newton, alchemist, cultist and deeply religious, were a brand new view of the man. We take this to mean that Burtt's view is now fully accepted and the appropriate revisions to history have been made or will soon be made in the textbooks. Addison-Wesley has a textbook division.

[91] H. McLachlan, *The Religious Opinions of Milton, Locke and Newton*, (New York: Russell and Russell, 1941); McLachlan, *Sir Isaac Newton Theological Manuscripts*, (Liverpool: University Press, 1950). See J. E.. McGuire and P.M. Rattansi, *Newton and the Pipes of Pan, Notes and Records of the Royal Society*, 21 (1966): 108-143; Frank E. Manuel, *A Portrait of Isaac Newton*, (Cambridge, Mass.,: Harvard Press, 1968); B.J.T. Dobbs, *The Foundations of Newton's Alchemy*, (Cambridge: Harvard Press, 1975); Richard S. Westfall, *Never At Rest: A Biography of Isaac Newton*, (Cambridge: Harvard Press, 1980); Scott Mandelbrote, *'A duty of the greatest moment': Isaac Newton and the writing of biblical criticism, British Journal of the History of Science*, 26, (1993): 281-302 and Michael White, *Isaac Newton, The Last Sorcerer*, (Menlo Park, California, Addison Wesley Co., 1998).

For a series of articles with a different point of view see: William H. Austin, *Isaac Newton on Science and Religion, Journal for the History of Ideas*, volume 31, (1970): 521-542; James W. Garrison, *Newton and the Relations of Mathematics to Natural Philosophy, Journal for the History of Ideas*, Volume 47, number 4, (1987): 609-607-627; and three other articles in the same volume celebrating Newton's Tercentenary. See also A Rupert Hall's latest books, discussed below.

[92]David Kubrin, Newton and the Cyclical Cosmos: *Providence and the Mechanical Philosophy, Journal of the History of Ideas*, 28 (1967): 325-346 and M.A. Hoskin, *Newton, providence and the universe, Journal for the History of Astronomy*, 8, (1977): 77-101.

Chapter 10

[1]See H. Floris Cohen, *The Scientific Revolution: A Historiographical Inquiry*, (Chicago: University of Chicago Press, 1994).

[2] Morton White, *Anti-intellectualism in America* in *Pragmatism and the American Mind*, (New York: Oxford University Press, 1973), 78-92.

[3] Ibid., 79.

[4] Ibid., 82.

[5] Ibid., 85.

[6] Ibid., 85.

[7]Stephen Toulmin, *From Form to Function: Philosophy and History of Science in the 1950s and Now, Daedalus* Volume I, number 106, (Summer, 1977): 143-162.

[8]Ibid., 145. Wittgenstein and Bertrand Russell were at Cambridge, a fact which might tend to qualify Toulmin's statements. Perhaps he means that the greater contribution to fashion was made by the German philosophers who came to the U.S. and that in Britain philosophy remained more balanced.

[9]Ibid., 146.

[10]Ibid., 146-147.

[11]Ibid., 148-149.

[12]Ibid., 150-151.

[13]Ibid., 154.

[14] Editions of *Principles and Problems of Right Thinking,* (Harper and Brothers) appeared in 1928, 1931 and 1946. The 1946 edition was re-titled, *Right Thinking: A Study of Its Principles and Methods.* The post war edition marked a solid break with the earlier ones. It include, for one thing, a section detailing Burtt's own theory of value and also his theory of philosophical method (Part IV).

[15] J.H. Randall, "The Development of Scientific Method in the School of Padua", *The Journal of the History of Ideas,* volume 1, number 2 (April, 1940) and *The School of Padua and the Emergence of Modern Science,* (Padua: Antenore, 1961).

[16] J.H. Randall, *The School of Padua and the Emergence of Modern Science,* (Padua: Antenore, 1961), 25.

17Burtt, *Right Thinking: A Study of Its Principles and Methods,* 382.

[18]John P. Anton, editors Preface to *naturalism and Historical Understanding, Essays on the Philosophy of John Herman Randall, Jr.,* (New York: State University Press, 1967).

Chapter 11

[1]Burtt, *Does Humanism Understand Man?, The Humanist,* Volume 5 (Autumn, 1945):114.

[2] Burtt, *Humanism and the Doctrine of Sin, The Humanist.* Vol. 5, (Winter, 1945).

[3] John H. Randall, *To Win Out, Must Humanists Embrace Sin? , The Humanist,* Vol. 6, number 1 (Spring, 1946).

[4]Charles Lyttle, Unitarian historian who, in 1952, chronicled the events of developing humanism in the United States. Quoted from Lyttle, *Freedom*

Moves West, (Boston, MA: Beacon Press, 1952) by Wilson, in Wilson, Preface, page 4.

[5] Burtt, *How Are Theological Conclusions Demonstrable?, The Journal of Liberal Religion* Volume 5, (Winter, 1944): 140-141.

[6] Ibid., 140.

[7]Corliss Lamont, *Voice In the Wilderness,* (Buffalo, NY.: Prometheus Books, 1974), 11.

[8]Ibid. xii.

[9]Point fourteen read: The humanists are firmly convinced that existing acquisitive and profit-motivated society has shown itself to be inadequate and that a radical change in methods, control, and motives must be instituted. A socialized and cooperative economic order must be established to the end that the equitable distribution of the means of life be possible. The goal of humanism is a free and universal society in which people voluntarily and intelligently cooperate for the common good. Humanists demand a shared life in a shared world.

[10]Wilson, Chapter 17, page 9.

[11]In 1973, twenty years after the symposium, a second Manifesto was circulated. Burtt did not sign the second *Manifesto* (1973), although Randall did.

[12]*The Humanist,* Volume 13, no. 2, (March/April, 1953): 63-71.

[13] Wilson, Chapter 17, page 2.

[14]*Principles and Problems of Right Thinking,* (New York and London: Harper and Brothers, 1928, 1931) and revised as *Right Thinking, a Study of Its Principles and Methods* (1946); and *The Problem of Philosophic Method, Philosophical Review,* volume 55, (1946): 505-533; and *The Problem of Theological Method, The Journal of Religion,* Volume XXVII, no. 1 (January, 1947): 1-26.
Wilson, Chapter 17, page 9.

[15]Arthur E. Murphy was Burtt's colleague during the Chicago crisis in the department of philosophy, who later became his department chairman at Cornell, hiring analytic philosophers to replace the idealistic "Hegelians."

[16]Wilson, Chapter 17, page 1.

[17]Ibid.

[18]Wilson, Chapter 18, page 2.

[19]Professor Stuart Brown, retired, professor emeritus from Cornell, who was both Burtt's student in the 1930s at Cornell, later, his teaching assistant, and finally, the department chairperson after Arthur Murphy, offered some candid insight, while qualifying his own feelings for Burtt. Quoting Professor Brown in a telephone conversation, "I really couldn't talk to him later on. He changed. He used to be a hard-headed philosopher. He used to love to go into a classroom and discuss logic, but later...(unfinished sentence) I think our relations were always good. I hope I was thought of as, at least, congenial." In

the same telephone interview, Mrs. Brown, on another line, remembered Burtt fondly. She said, "After Marjorie [Burtt's second wife] entered Ned's life he became almost 'other worldly', more human, easier to talk to." She said that earlier on Burtt had been "impossible to live with" and explained something of the marital discord that had troubled Burtt enough to enter the long period of psychoanalysis. The Browns thought that Burtt had undergone some kind of "conversion." "He changed so very radically, after the divorce," Professor Brown went on. "Sometime during the late 1930s Ned declared that his philosophical leaning had been a mistake." Mrs. Brown added, "Personally, Ned became more relaxed at that time." Sometime during the 1930s, Burtt had become a Quaker. A family member remarked that Burtt's professional stature was seen to have suffered because of the stigma of divorce, not to mention the emotional loss, since Burtt no longer lived with his four daughters and lost his family life.

20 Burtt, *Types of Religious Philosophy*, revised edition, (New York: Harper and Brothers, 1951), 440-442.

21 Personal correspondence, February 4, 1997.

22 Burtt, *Types of Religious Philosophy*, 342.

23 Ibid., 343. There is no doubt that Burtt is putting forward his own opinion, since he cites *Right Thinking*, Chapters III and IV.

24*The Problem of Philosophic Method, Philosophical Review*, Vol. 55, (Sept., 1947): 505-533 ; *The Problem of Theological Method, Journal of Religion*, Vol. 27 (Jan., 1947): 1-15, and*Right Thinking, A Study of Its Principles and Methods.* New York and London: Harper & Brothers, 1928, 1931 published as *Principles and Problems of Right Thinking* and revised significantly in 1946.

25Burtt, *In Search of Philosophic Understanding*, (Indianapolis and Cambridge: Hackett Publishing Company), 1965, reprinted 1980.

26John E. Smith, *Review of* In Search of Philosophic Understanding, , *Philosophical Review*, 78 (Jan. 1969): 99-102.

27 Burtt, *Right Thinking, Preface*, ix

28 Ibid., 626.

29 Ibid., 658.

30Ibid., 660-661.

31Ibid., 725

32Ibid., 642.

33Ibid., 665.

34Ibid., 711-712.

35Ibid., 714.

36 Summarized from Ibid., 715-730.

37Burtt, *My Path to Philosophy*, 430.

[38]Ibid., 430-431.

[39] In 1957 Burtt's *The Value Presuppositions of Science* appeared in *The Bulletin of Atomic Scientists,* Vol. 13, number 3 (March, 1957): 99-106, reprinted in *The New Scientist: Essays on the Methods and Values of Modern Science,* Paul Obler and Herman A. Estrin, editors, (New York: Anchor Books, 1962). By then Burtt had retired and of course, Kuhn's *Structure of Scientific Revolutions* published shortly thereafter (1962).

Chapter 12

[1]Hollinger, *T.S. Kuhn's Theory of Science and Its Implications for History, American Historical Review,* 78, (1973): 370.

[2]Collingwood, *Essay on Metaphysics,* (Oxford: Clarendon Press, 1940), 74. This passage was type-copied by Burtt for his own study.

[3]Ibid., 74-77.

[4]Ibid.

[5]*In Search of Philosophic Understanding,* footnote, 130.

[6]Burtt, *The Philosophy of Man as All-embracing Philosophy, The Philosophical Forum,* Volume II, number 2 (New Series), (Winter, 1970-1971): 159-171. Also delivered as an address to the Eastern Division of the American Philosophical Association in 1969.

[7]*The Status of World Hypotheses* Discussion *The Philosophical Review.* Volume 52 (November, 1943).

[8]*The Problem of Philosophic Method,* footnote, 520.

[9]Ibid. Burtt had recommended Sheldon's *Philosophy's Job Today,* (Berkeley and Los Angeles: University of California Press, 1947) to the seminar group as well as Pepper's *World Hypotheses.*

[10]Burtt, *The Generic Definition of Philosophic Terms. The Philosophical Review,* Volume 62, (January, 1953), 57. Also read before the philosophical club at Cornell on November 29, 1951 and before the Philosophical Forum of Princeton University, December 4, 1951.

[11]Burtt, *The Generic Definition of Philosophic Terms, The Philosophical Review,* Volume LXII, No. 1, (January, 1953): 42.

[12]Personal correspondence, 8 January, 1996.

[13]Wilmon H. Sheldon, *America's Progressive Philosophy,* (New Haven, Connecticut: Yale University Press, 1942), 73.

[14]Ibid., 165.

[15]Ibid., 232

[16]Dorothy M. Emmet, *The Nature of Metaphysical Thinking,* (London: Macmillan and Co. Ltd and New York: St Martin Press 1951), 91.

[17]Albert Hofstadter, *A Conception of Empirical Metaphysics, The Journal of Philosophy*, Volume XLV, No. 16, (July 19, 1948): 433.

[18]Ibid., 434.

[19]E.T. Owen, *The Illusion of Thought, The Journal of Philosophy*, Volume XLV, No. 19 (September 9, 1948), 510.

[20]Burtt, *Descriptive Metaphysics, Mind,* 72 (January, 1963), 34-35.

[21]Ibid., 37.

[22]Ibid., 38.

[23]Bernard Wand, another graduate student at Cornell in the late 1940s, uses these words to describe the "tension" between Black and Burtt and notes that Black's own Critical Thinking was being taught by him at the University. Wand to Villemaire, December 19, 1995.

[24]Collingwood, 40-43.

[25]Burtt, *My Path to Philosophy*, 432.

[26]Ibid.

[27]Burtt, *The Human Journey*, 113.

[28]Ibid., 116.

[29]Ibid., 103.

[30]Burtt had completed all the degree requirements in 1923, but had to wait for his degree because he did not have enough money to have his dissertation copied 100 times for the Columbia library. Publication in 1925 in the U.S. satisfied this requirement.

[31] See *The Human Journey*--Stephanos Nirmalendu Ghosh Lectures on Comparative Religion 1968, (Calcutta, India: University of Calcutta, 1981)

[32]*In Search of Philosophic Understanding*, 1965. John E. Smith reviewed this book in 1969, *Philosophical Review* 78 (January, 1969): 99-102.

[33] Burtt, *Man Seeks the Divine*, 454-455.

[34]The Festschrift was never completed. Any contributions, besides Hackett's, were returned to their authors by Lindley.

[35] Burtt to Stuart Hackett, January 18, 1971, Rare and Manuscript Collections Carl A. Kroch Library, Cornell University. Burtt Archive Box.

[36] Stuart M. Brown and Nicholas Sturgeon, *Eulogy for Edwin Arthur Burtt*, corrected copy dated October 29, 1990. Photocopy from Professor Sturgeon's file, page 5.

[37] Elsie Myers Stainton, *E.A. Burtt Bibliography* reprint from *Philosophy East & West*, Volume 22, Number 4 (October 1972). Other short articles, mostly unpublished or privately published are located in the Burtt Archive Boxes at Olin Library, Carl A. Kroch Library, Rare and Manuscript Collection, Cornell University, Ithaca, NY.

[38]Burtt, *Preface* to *Philosophy: Yesterday, Today and Tomorrow,* unpublished manuscript, pages 1-2, rare and manuscript collections, Carl A. Kroch Library, Cornell University, Ithaca, NY. Burtt Archive Box. The library manuscript is incomplete. Burtt's daughter, Freddie Burtt-Brinster of Seattle, Washington has the final complete draft of the book.

[39]Burtt's last writing, including a book, privately published, *Light, Love and Life* (1985) and a number of short pieces written for Quaker "sermons" or "talks" concentrate on God, Spiritual Growth, Love and Mystical Wisdom.

Notes to the Conclusion

[1] Burtt, My Path to Philosophy, 436-437.

[2] Francis Moriarty, *The Philosophy of E.A. Burtt,* Ph.D. dissertation, University of Adelaide, Politics Department, 1994, 37. Some of these letters are in archive boxes at the Carl Kroch library, Rare and Manuscript Collections, Cornell University, but most correspondence information is secondary, word of mouth reporting.

[3]Correspondence Burtt to Hackett, December 1, 1970 and January 18, 1971 Archive Box #1, #14/21/2593 located in Rare and Manuscript Collections, Carl A. Kroch Library, Cornell University, Ithaca, New York.

[4] The Human Journey appeared in 1981, University of Calcutta Press, Calcutta, India.

[5]Bertocci, Peter A., editor, *Mid-Twentieth Century American Philosophy: Personal Statements,* (New York: Humanities Press, 1974), 104-119.

[6] Burtt, *Toward a Philosophy of Philosophy* in Ibid., 112-113.

[7] Ibid., 116-118.

[8] Peter Bertocci, *Love and Reality In E.A. Burtt's Philosophy: A Personalistic Critique*, Reprinted from *Idealistic Studies*, Vol. 5, No. 3 (September, 1975): 269-289. My copy from Archive Box #1, #14/21/2593, Rare and Manuscript Collections, Carl Kroch Memorial Library, Cornell University, Ithaca, New York.

[9]From hand written pages "Parts of Letters from Ned Burtt to Helen and Karl Schantz." Prepared by Helen Schantz for this project, 1994.

[10] Ralph B. Winn, editor, *American Philosophy,* (New York: Philosophical Library, Inc., 1955), see page 217. Occasionally Burtt is recognized as a Humanist philosopher in works on Humanism, stemming from his having signed the *Humanist Manifesto I.* In the main, works on American philosophers have ignored Burtt.

[11]Stratford Caldecott to Edwin Burtt, 16 February, 1987, Archive Box #1, #14/21/2593 located in Rare and Manuscript Collections, Carl A. Kroch Library, Cornell University, Ithaca, New York.

[12] These books are *The Metaphysical Foundations of Modern Physical Science: A Historical and Critical Essay,* (Kegan Paul, Trench, Trubner & Co., 1924 and 1932, revised); *Types of Religious Philosophy,* (Harper and Brothers, 1939 and 1951, revised); *The Teachings of the Compassionate Buddha,* (Mentor Books, 1955) and *Man Seeks the Divine: A Study of the History and Comparison of Religions,* (Harper and Brothers, 1957).

[13]Karl C. Schantz, Ned Burtt, private testimonial and remembrance, typed carbon copy, given to me by Karl Schantz.

[14] Thomas Kuhn, *The Structure of Scientific Revolutions,* Chicago: University of Chicago Press, 1962), 153 and Richard Rorty, *Philosophy and the Mirror of Nature,* (London: Basil Blackwell, 1986), 65.

[15] Burtt, *In Search of Philosophic Understanding,* New American Library, 1965 and Indianapolis, Indiana: Hackett Publishing Company, 1980, 61-62.

[16]Burtt, *Toward a Philosophy of Philosophy* in Peter Bertocci, editor, *Mid-Twentieth Century American Philosophy: Personal Statements.* New York: Humanities Press, 1974, 115-116.

[17] Ibid., 118.

[18]Burtt, *My Path to Philosophy, Philosophy East and West,* Volume 22, Number 4 (October, 1972): 430.

[19] *Truth, Understanding, and Philosophy,* a Presidential address delivered before the Sixty-First Annual Meeting of the Eastern Division of the American Philosophical Association at Boston, Massachusetts, December 27-29, 1964 in *Proceedings and Addresses of The American Philosophical Association.* Volume XXXVIII (October,1965). The Antioch Press, 10-11. See also *In Search of Philosophic Understanding,* (Indianapolis and Cambridge: Hackett Publishing Company, 1965, reprinted 1980).

[20] Edwin A. Burtt, "Toward a Philosophy of Philosophy" in *Mid-Twentieth Century American Philosophy: Personal Statements* edited by Peter A. Bertocii, (New York: Humanities Press, 1974), 104.

[21]Burtt, *Metaphysics of Empirical Theology,* a mimeographed typewritten document, with notations in Burtt's handwriting, found among his papers at 277 North Willard Way in Ithaca, New York.

[22]Burtt, *Toward a Philosophy of Philosophy,* 104.

[23]Ibid., 106.

BIBLIOGRAPHY

Adler, Mortimer J. Philosopher at Large: An Intellectual Autobiography. New York: Macmillan, 1977.

Anton, John P. *John Herman Randall, Jr., A Biographical Sketch* in John Anton, editor. Naturalism and Historical Understanding, Essays on the Philosophy of John Herman Randall, Jr. Albany, New York: State University of New York Press, 1967, vii-x.

Austen, Willard. *Building Without An Architect, How we built a ten-room English cottage of stone and wood.* Country Life In America (October, 1903): 431-433.

Barnes, Harry Elmer. *The Historian and the History of Science.* Scientific Monthly Volume 11 (1919): 112-126.

____. *Does Science Require a New Conception of God?* Current History, XXIX, number 6, (March, 1929): 883-898.

____. *John Herman Randall, Jr.: A Memoir and a Tribute* in John P. Anton, editor. Naturalism and Historical Understanding, Essays on the Philosophy of John Herman Randall, Jr. Albany, New York: State University of New York Press, 1967, 277-281.

Bennett, Charles A. *The Paradox of Mysticism.* Proceedings of the Sixth International Congress of Philosophy. Harvard University, Cambridge, Massachusetts, September 13, 14, 15, 16, 17, 1926. Edgar Sheffield Brightman, editor. New York and London: Longmans, Green, 1927.

Bernstein, Richard J. *John Dewey* in The Encyclopedia of Philosophy. Paul Edwards, editor, New York: Collier Macmillan Publishers, 1967.

Bertocci, Peter A., editor. Mid-Twentieth Century American Philosophy: Personal Statements. New York: Humanities Press, 1974.

____. *Love and Reality In E.A. Burtt's Philosophy: A Personalistic Critique.* Reprinted from Idealistic Studies, Vol. 5, No. 3 September, 1975. My copy from Archive Box #1, #14/21/2593, Rare and Manuscript Collections, Carl Kroch Memorial Library, Cornell University, Ithaca, New York.

Boulding, Kenneth E. The Meaning of the Twentieth Century, The Great Transition. Part of the world Perspectives Series, Volume 34. New York and London: Harper and Row, 1964.

Bourne, Randolph S. War and the Intellectuals, Collected Essays, 1915-1919, Carl Resek, editor. New York: Harper and Row, 1964.

Braybrooke, David. Seminar Notes from E. A. Burtt's Graduate Level Course, Philosophy 587, Collingwood's Metaphysics. Taught at Cornell University, 1950-51.

Brown, Stuart, and Sturgeon, Neil. 1990. *Eulogy for Edwin Arthur Burtt.* Cornell University: Photocopy.

Buchdalh, Gerd. Metaphysics and the Age of Science, The Classical Origins: Descartes to Kant. Cambridge, MA: MIT Press, 1969

Butterfield, Herbert. The Origins of Modern Science 1300-1800. London: 1949 and New York: The Free Press, 1957, revised 1965.

Butts, Robert E. *Whewell on Newton's Rules of Philosophizing* in The Methodological Heritage of Newton, Robert Butts and John W. Davis, editors. Toronto: University of Toronto Press, 1970.

Cahoone, Lawrence, Editor. From Modernism to Postmodernism, An Anthology. Cambridge, Massachusetts: Blackwell Publishing Co., 1996.

Campbell, James. Understanding John Dewey. Chicago and LaSalle, Illinois: Open Court, 1995.

Cohen, H. Floris. The Scientific Revolution: A Historiographical Inquiry. Chicago: University of Chicago Press, 1994.

Cohen, I. Bernard. *Alexandre Koyré (1892-1964) Commemoration.* Isis. 57 (1966): 157-165.

Cohen, Morris R. *Why I Am Not a Communist* in *A Symposium on Communism.* The Modern Monthly. Volume VIII, Number 3, (April, 1934): 138-142.

____. A Dreamer's Journey, The Autobiography of Morris Raphael Cohen. Boston, Massachusetts: Beacon Press, 1949.

____. American Thought, A Critical Sketch. Glencoe, Illinois: The Free Press, 1954.

Collingwood, R. G. Essay on Metaphysics. Cambridge, England: Clarendon Press, 1940.

Columbia Associates in Philosophy. An Introduction to Reflective Thinking. Boston, New York, Chicago, San Francisco: Houghton Mifflin, 1923.

Cook, Gary A. George Herbert Mead, The Making of a Social Pragmatist. Urbana and Chicago: University of Illinois Press, 1993.

Croce, Paul Jerome. Science and Religion In the Era of William James. Eclipse of Certainty 1820-1880. Chapel Hill, NC: University of North Carolina Press, 1995.

Cornell Daily Sun, Editorial Staff. *Far From Unamerican.* (Monday, April 9, 1951). Reprinted from Frank Moriarty, 1994.

Culler, Jonathan. On Deconstruction: Theory and Criticism after Structuralism. Ithaca: Cornell University Press, 1982.

Cunningham, Andrew. *Getting the Game Right: Some Plain Words on the Identity and Invention of Science.* Studies in the History and Philosophy of Science. 19 (1988): 365-389.

____. and Williams, Perry. *De-centering the 'big picture': The Origins of Modern Science and the modern origins of science.* British Journal for the History of Science. 26 (1993): 407-432.

Daston, Lorraine. *History of Science in an Elegiac Mode, E.A. Burtt's Metaphysical Foundations of Modern Physical Science Revisited.* Isis. 82 (1991): 522-531.

Delaney, C.F. Mind and Nature, A Study of the Naturalistic Philosophy of Cohen, Woodbridge and Sellars. London: University of Notre Dame Press, 1969.

Dewey, John. Reconstruction in Philosophy. New York: Henry Holt and Company, 1920.

____. How We Think. Boston, New York and Chicago: D.C. Heath, 1910.

____. *Our National Dilemma.* New Republic. XXII (March 24, 1920): 118.

____. Types of Thinking including A Survey of Greek Philosophy, Robert W. Clopton and Tsuin-Chen Ou, eds. and translators, *Introduction* by Samuel Meyer. New York: Philosophic Library, 1983.

Diggins, John. The Promise of Pragmatism And the Crisis of Knowledge and Authority. Chicago: University of Chicago Press, 1994.

Donaldson, Scott. Archibald MacLeish, An American Life. New York and London: Houghton Mifflin Co., 1992.

Duhem, Pierre. The Aim and Structure of Physical Theory. New York: Athenaeum Books, 1977. First published 1906.

Durbin, Paul T., editor. A Guide to the Culture of Science, Technology, and Medicine. New York and London: The Free Press and Collier Macmillan, 1984.

Eames, Morris S. Pragmatic Naturalism, an Introduction. Carbondale and Edwardsville, London and Amsterdam: Southern Illinois University Press and Feffer & Simons, Inc., 1977.

Edman, Irwin. John Dewey, His Contribution to the American Tradition. New York: Greenwood Press, 1968.

Edwards, Paul, editor. The Encyclopedia of Philosophy. New York: Macmillan Co., reprint edition, 1972.

Emmet, Dorothy M. The Nature of Metaphysical Thinking. (London and New York: Macmillan and Co. Ltd. and St. Martin Press, 1961)

Fireside, Harvey. *Ned As We Knew Him.* The Ithaca Journal. Date unknown. Photocopy from the files of Helen and Karl Schantz.

Frankel, Charles, editor. The Golden Age of American Philosophy. New York: George Braziller, Inc., 1960.

Freeman, Joseph. An American Testament, A Narrative of Rebels and Romantics. New York: Octagon Books, 1973.

Force, James E. *Hume's Interest in Newton and Science* in James E. Force and Richard H. Popkin, editors. Essays on the Context, Nature, and Influence of Isaac Newton's Theology. Dordrecth, Boston and London: Kluwer Academic Publishers, 1990, 181-206.

Gaenslan, Tony. *Reminiscences of Ned Burtt.* Photocopied transcript from a tape, prepared by Freddie Burtt Brinster, 1989.

Gillott, John and Kumar, Manjit. Science and the Retreat From Reason. New York: Monthly Review Press, 1997.

Ginev, Dimiter. *Scientific Progress and the Hermeneutic Circle.* Studies In the History and Philosophy of Science. Volume 19, no. 3 (1988): 391- 395

Greenstreet, W. J., Editor. Isaac Newton 1642-1727, A Memorial Volume. The Mathematical Association. London: G. Bell and Sons, Ltd., 1927.

Guerlac, Henry. Essays and Papers in the History of Modern Science. Baltimore: Johns Hopkins University Press, 1977.

Gutmann, James. *John Herman Randall, Jr.: A Memoir of his Career at Columbia, 1915-1967* in John P. Anton, editor. Naturalism and Historical Understanding, Essays on the Philosophy of John Herman Randall, Jr. Albany, New York: State University of New York Press, 1967, 283-287.

Hall, A. Rupert. *Recollection & Reflections, Beginnings In Cambridge.* Isis. 75 (1984): 22-25.

Hartshorne, Charles. *Metaphysics for Positivists.* Isis. 2 (July, 1935): 287-303.

Hatfield, Gary. *Metaphysics and the New Science* in David C. Lindberg and Robert S. Westman, editors, Reappraisals of the Scientific Revolution. Cambridge: Cambridge University Press, 1990, 93-166.

Hesse, Mary. Revolutions and Reconstructions In the Philosophy of Science. Bloomington, Indiana: Indiana University Press, 1980.

Hofstadter, Albert. *A Conception of Empirical Metaphysics.* The Journal of Philosophy. Volume XLV, no. 16 (July 29, 1948): 421-435.

Hollinger, David A. *T.S. Kuhn's Theory of Science and Its Implications for History.* American Historical Review. 78 (1973): 370-393.

____. Morris R. Cohen and the Scientific Ideal. Cambridge, Massachusetts, MIT Press, 1975.

____. *Ethnic Diversity, Cosmopolitanism and the Emergence of the American Liberal Intelligentsia.* American Quarterly, XXVII (1975): 133-151.

____. *The Problem of Pragmatism in American History.* Journal of American History. 67 Number 1 (June, 1980): 88-107.

Hook, Sidney. *Why I Am a Communist, Communism Without Dogmas* in *A Symposium on Communism.* The Modern Monthly. Volume VIII, Number 3, (April, 1934): 143-165.

____. Out of Step, an Unquiet Life in the Twentieth Century. New York: Harper & Row, 1955.

Hooykaas, R. Religion and the Rise of Modern Science. Edinburgh: Scottish Academic Press, Ltd., 1972.

Hoyningen-Huene, Paul. Reconstructing Scientific Revolutions, Thomas S. Kuhn's Philosophy of Science. Translated by Alexander T. Levine with a Foreword by Thomas S. Kuhn. Chicago and London: University of Chicago Press, 1993.

Humanist Manifestos I and II. Buffalo, NY.: Prometheus Books, 1973. *Humanist Manifesto I* first appeared in The New Humanist. Vol. VI, number 3, (May/June 1933). *Humanist Manifesto II* first appeared in The Humanist. Vol. XXXIII, number 5. (September/October 1973).

Hume, David. *Concerning Natural Religion* in Philosophical Works, ed. by T. H. Green and T. H. Grose, London, 1874 and in The English Philosophers from Bacon to Mill, edited and with an *Introduction* by E.A. Burtt. New York: Modern Library, Random House, 1939.

Hurlbutt, Robert H. III. Hume, Newton and The Design Argument. Lincoln, Nebraska and London: University of Nebraska Press., 1963 and 1985.

____. *David Hume and Scientific Theism.* Journal of the History of Ideas. Volume XVII, number 4. (October, 1956): 486-497.

Jones, Howard Mumford. American Humanism, Its Means for World Survival. Westport, Connecticut: Greenwood Press, 1957, reprinted 1972. Part of the World Perspectives Serives, Volume 14.

Jones, William Frank. Nature and Natural Science, The Philosophy of FJE Woodbridge. Buffalo, NY.: Prometheus Books, 1983.

Kallen, Horace, M. and Hook, Sidney. American Philosophy Today and Tomorrow. New York: Lee Furman, Inc., 1935.

Keynes, John Maynard. *Newton, the Man* in The Collected Writings of John Maynard Keynes. Volume X of Essays in Biography. New York: Macmillan, St. Martin's Press for the Royal Economic Society, 1972, first edition 1933. The paper was read by Sir Geoffrey Keynes at the Newton Tercentenary Celebrations at Trinity College Cambridge, on 17 July 1946, not revised by its author, who had written it some years earlier.

Konvitz, Milton. *In Honor of E. A. Burtt.* Philosophical Review. (October, 1972): vii. (The issue was dedicated to E. A. Burtt in honor of his eightieth birthday.)

____. *E.A. Burtt (1892-1989).* Photocopy, with hand written corrections, from the files of Milton Konvitz.

Koestler, Arthur. The Sleepwalkers, A History of Man's Changing Vision of the Universe, with an Introduction by Herbert Butterfield. London and New York: Penguin Books, first published 1959, reprint edition, 1989.

Krikorian, Yervant H. Recent Perspectives in American Philosophy. The Hague, Netherlands: Martinus Nijhoff, 1973.

_____, Editor. Naturalism and the Human Spirit. Morningside Heights, NY: Columbia University Press, 1994.

Kublick, Bruce. The Rise of American Philosophy. New Haven, Connecticut and London: Yale University Press, 1977.

Kuhn, Thomas S. The Copernican Revolution, Planetary Astronomy in the Development of Western Thought. Cambridge and London: Harvard University Press, 1957.

____. The Structure of Scientific Revolutions, second edition. Chicago: The University of Chicago Press, 1962, 1970.

____. *Alexandre Koyre & the History of Science, On an Intellectual Revolution.* Encounter. (Jan., 1970): 67-69.

____. *Recollection & Reflections, Professionalization Recollected In Tranquillity.* Isis. 75 (1984): 28-32.

Kurtz, Paul., Editor. Philosophic Essays in Pragmatic Naturalism. Buffalo, NY: Prometheus Books, 1990.

____, editor. Sidney Hook, Philosopher of Democracy and Humanism. Buffalo, NY: Prometheus Books, 1983.

____. American Philosophy in the Twentieth Century, A Sourcebook from Pragmatism to Philosophical Analysis. New York and London: Macmillan Co., 1966 and 1968.

Lakatos, Imre and Alan Musgrave, Editors. Criticism and the Growth of Knowledge, Proceedings of the International Colloquium in the Philosophy of Science, 1965 Volume 4. Cambridge: Cambridge University Press, 1970.

Laing, Dean G.J. *The Situation in the Department of Philosophy.* University of Chicago Manuscript Collection. The Faculty of the Division of the Humanities: December 1930-October, 1940, Kept in the Secretary of the Faculties' Office at the University of Chicago, filed under 12/3/1931. Presidents' Papers', ca. 1925-1945, University of Chicago, Regenstein Library, Department of Special Collections, Chicago, Ill.

Lamont, Corliss. Voice In the Wilderness. Buffalo, NY.: Prometheus Books, 1974.

____. The Philosophy of Humanism. New York: The Half-Moon Foundation, 1990.

Lasch, Christopher. The New Radicalism in America 1889-1963. The Intellectual as a Social Type. New York: A.A. Knoph, 1965.

Libby, Walter. An Introduction to the History of Science. Boston: Houghton Mifflin, 1917.

Lindberg, David. *Introduction* to Reappraisals of the Scientific Revolution, David C Lindberg and Robert S. Westman, editors. Cambridge and New York: Cambridge University Press, 1990.

____. The Beginnings of Western Science. Chicago and London: University of Chicago Press, 1992.

Lippmann, Walter. *Uneasy America.* New Republic. V (December 25, 1915): 195-196.

Mach, Ernst. Science of Mechanics. London: Kegan Paul, Trench, Trubner & Co. and Chicago: Open Court, 1907

Magill, Frank N. and McGreal, Ian P., editors. Masterpieces of World Philosophy in Summary Form. New York: Harper and Brother, 1961.

Matlack, Jean. *A Life Study of a Psychotherapist: Marjorie Murray Burtt.* Ph.D. dissertation in Education at the University of Massachusetts, 1976.

May, Henry F. The End of Innocence, A Study of the First Years of Our Own Time, 1912-1917. New York: Alfred A. Knopf, 1959.

____. The Discontent of the Intellectuals: A Problem of the Twenties. Chicago: Rand McNally, 1963.

McDougall, Walter A. The Heavens and the Earth, A Political History of the Space Age. New York: Basic Books, 1985.

McNeill, William H. Hutchins' University, A Memoir of the University of Chicago 1929-1950. Chicago and London: University of Chicago Press, 1991.

Meade, George Herbert. The Philosophy of the Present. LaSalle, Illinois: Open Court, 1959.

Meyer, Samuel, editor. Types of Thinking including A Survey of Greek Philosophy by John Dewey. New York: Philosophical Library, 1984.

Moriarty, Francis. *The Philosophy of E. A. Burtt, The Metaphysical Foundations for a World Community.* Ph.D. dissertation in Political Science at the University of Adelaide, Australia, 1994.

Murphy, Arthur E. Reason, Reality, and Speculative Philosophy, Marcus Singer, editor, with an Introduction by Marcus Singer, Madison, Wisconsin: University of Wisconsin Press, 1996.

Murphy, John P. Pragmatism From Peirce to Davison with an Introduction by Richard Rorty. Boulder, San Francisco and Oxford: Westview Press, 1990.

New Republic, editorial staff. *Pacifism vs. Passivism*. I (December 12, 1914): 6-7.

____. *War at Any Price*. V (Nov. 27, 1915).

____. *An Inconclusive Peace*. III (July 24, 1915).

____. *Aggressive Pacifism*. V (January 15, 1916): 263-265.

Noble, David F. America by Design, Science Technology and the Rise of Corporate Capitalism. New York: Alfred A. Knopf, 1982.

____. Forces of Production, A Social History of Industrial Automation. New York: Alfred A. Knopf, 1984.

Owen, E. T. *The Illusion of Thought*. The Journal of Philosophy. Volume XLV, number 19 (September 9, 1948): 505-511.

Passmore, John. A Hundred Years of Philosophy. New York: Macmillan Company, 1957.

Penick, James L. Jr., et. al., editors, The Politics of American Science 1939 to the Present. Cambridge, Massachusetts, and London: MIT Press, 1965 and 1972.

Pepper, Stephen. World Hypotheses: A Study in Evidence. Berkeley and Los Angeles: University of California Press, 1948.

Pyun, Hae Soo. Nature, Intelligibility, and Metaphysics, Studies in the Philosophy of F. J. E. Woodbridge. Amsterdam: B.R. Gruner N.V., 1972.

Randall, John Herman, Jr. The Making of the Modern Mind. New York: Houghton Mifflin Co., 1926.

____. *Effects of Science and Industry on Religion*. Current History. XXX Number 3 (June, 1929): 355-366.

____. *The Development of Scientific Method in the School of Padua*. The Journal of the History of Ideas. Volume 1, number 2, (April, 1940).

____. *Dean Woodbridge*. Columbia University Quarterly. Volume 32, 1940.

____. *Newton's Natural Philosophy, Its Problems and Consequences* in F. P. Clarke and M. C. Nahm, editors. Philosophical Essays in Honor of Edgar Arthur Singer, Jr. Philadelphia, PA: University of Pennsylvania Press and London: Humphrey Milford, Oxford University Press, 1942, pages 335-357.

____. Theory and Practice in Historical Study: A Report of the Committee on Historiography. New York: Social Science Research Council Bulletin 54, 1946.

____. The School of Padua and the Emergence of Modern Science. Padua: Antenore, 1961.

____. *To Win Out, Must Humanists Embrace Sin?*. The Humanist. Volume 6, number 1, (1946): 20-27.

____. *The Place of Leonardo Da Vinci in the Emergence of Modern Science*. Journal of the History of Ideas. Volume XIV, number 2 (April, 1953): 191-202.

____. How Philosophy Uses Its Past. New York: Columbia University Press, 1963.

____. Autobiographical sketch preceding the essay, *Historical Naturalism* in Horace M. Kallan and Sidney Hook, editors. American Philosophy Today and Tomorrow. New York: Lee Furman, Inc., 1935.

____. *Towards a Functional Naturalism* in John E. Smith, editor. Contemporary American Philosophy, Second Series. London and New York: George Allen Unwin and Humanities Press, 1970.

____. Philosophy After Darwin Chapters for The Career of Philosophy Volume III and Other Essays, Beth Singer, editor. New York: Columbia University Press, 1977.

____ and J. Buchler. *The Department of Philosophy* in Randall, Hadas, Torrey, et. al., editors. A History of the Faculty of Philosophy Columbia University. New York: Columbia University Press, 1977.

Reck, Andrew J. The New American Philosophers. An Exploration of Thought Since World War II. Baton Rouge, LA: Louisiana State University Press, 1968.

Redondi, Pietro. *De l'histoire des sciences a l'histoire de las pensee scientifique: le combat d'Alexandre Koyre*. Preface to De la mystique a las science, cours, conferences et documents, 1922-1962. Paris: Editions de l'ecole des hautes etudes in sciences sociales, 1986.

Rescher, Nicholas. Process Metaphysics: An Introduction to Process Philosophy. State of New York Press, 1996.

Robinson, James Harvey. The Mind in the Making. New York and London: Harper and Brothers, 1921.

____. The New History. New York: Macmillan Co., 1912.

Rorty, Richard. Philosophy and the Mirror of Nature. London: Basil Blackwell, 1986.

____. Consequences of Pragmatism (Essays 1972-1980). Minneapolis, Minnesota: University of Minnesota Press, 1982.

____. *Knowledge and Acquaintance*. Book Review of Bertrand Russell: The Spirit of Solitude, 1872-1921 by Ray Monk. The New Republic, (December 2, 1996).

Rucker, Darnell. The Chicago Pragmatists. Minneapolis: U. of Minnesota Press, 1969.

Russell, Bertrand. *A Free Man's Worship* in Philosophical Essays. New York, Bombay, and Calcutta: Longmans, Green, and Co., 1910.

____. *Why I Am Not a Communist* in *A Symposium on Communism*. The Modern Monthly. Volume VIII, Number 3, (April, 1934): 133-137.

____. History of Western Philosophy and Its Connection with Political and Social Circumstances from the Earliest Times to the Present Day. London: George Allen and Unwin Ltd., first published in 1946, New edition (reset), 1961.

____. *Review of Metaphysical Foundations of Modern Physical Science.* Dial 79 (September, 1925): 255-258 and Nation and Athenaeum 37 (June 13, 1925): 326.

Ryder, John. American Philosophic Naturalism in the Twentieth Century. Amherst, New York: Prometheus Books, 1994.

Sarton, George. *L'Histoire de las Science.* Isis Number 1, Volume 1, (1913): 3-46.

____. The History of Science and the New Humanism. Cambridge, Massachusetts: Harvard University Press, 1937.

Schantz, Karl. *Ned Burtt.* Carbon copy from Karl Schantz's file.

Schlesinger, Arthur M., Jr. and White, Morton, editors. Paths of American Thought. Boston: Houghton Mifflin Company, 1963.

Schneider, Herbert W. A History of American Philosophy. New York: Columbia University Press, 1946.

Shapere, Dudley. Galileo, A Philosophical Study. Chicago and London: University of Chicago Press, 1974.

Shapin, Steven. The Scientific Revolution. Chicago and London: The University of Chicago Press, 1996.

Sheldon, Wilmon H. America's Progressive Philosophy. New Haven, Connecticut: Yale University Press, 1942.

____. Philosophy's Job Today. Berkeley and Los Angeles: University of California Press, 1947.

Sherman, Tamar Asedo. *On Changing Careers in Midstream.* Photocopy of an article on Marjorie Burtt from an unidentified Ithaca Newspaper. From the files of Professor Milton Konvitz.

Shotwell, James T. An Introduction to the History of History. New York: Columbia University Press, 1922.

____. The Religious Revolution of Today. Boston and New York: Houghton Mifflin Co., 1913.

____. On the Rim of the Abyss. New York: Macmillan, 1936.

Singer, Marcus G. *Two American Philosophers: Morris Cohen and Arthur Murphy* in Marcus G. Singer, editor. American Philosophy. Cambridge and London: Cambridge University Press, 1985, 295-329.

____. *The Context of American Philosophy* in Marcus G. Singer, editor. American Philosophy. Cambridge and London: Cambridge University Press, 1985, 1-20.

____. *Memoir* in Arthur E. Murphy, Reason, Reality, and Speculative Philosophy. Madison, WI: University of Wisconsin Press, 1996. Professor Singer provided me with an advance manuscript copy.

Stiles, Martin B. *Unit of Survival Now Mankind as a Whole, So writes philosopher Edwin A. Burtt, 90.* The Ithaca Journal. Date unknown. Photocopy from the files of Helen and Karl Schantz.

Smith, John E. *Review of In Search of Philosophic Understanding,*. Philosophical Review. 78 (Jan. 1969): 99-102.

Smith, A. Mark. *Galileo's Theory of Indivisibles: Revolution or Compromise?* Journal of the History of Ideas, XXXVII Number 4 (October-December, 1976) : 571-588.

Smith, T.V. A Non-Existent Man, An Autobiography. Austin, Texas: University of Texas Press, 1962.

Stainton, Elsie Myers. *E.A. Burtt Bibliography* reprint from Philosophy East & West. Volume 22, Number 4, (October 1972).

Straker, Stephen M. *What is the history of theories of perception the history of?* in Religion, Science and World view, Essays in Honor of Richard S. Westfall, edited by Margaret J. Osler and Paul Lawrence Farber. Cambridge: Cambridge University Press, 1985, pp. 245-273.

Stevenson, Louise L. Scholarly Means to Evangelical Ends. The New Haven Scholars and the Transformation of Higher Learning in America 1830-1890. Baltimore and London: Johns Hopkins Press, 1986.

Tejera, Victorino. American Modern, The Path Not Taken, Aesthetics, Metaphysics and Intellectual History in Classic American Philosophy. Lanham, Boulder, New York, London: Rowman & Littlefield Publishers, Inc., 1996.

Thackray, Arnold. *History of Science* in A Guide to The Culture of Science, Technology and Medicine. Edited by Paul T. Durbin. New York: The Free Press, 1984.

The Ithaca Journal. Local Obituaries, Edwin Arthur Burtt. Friday September 8, 1989. Photocopy from the files of Helen and Karl Schantz.

Thorndike, Lynn. The History of Magic and Experimental Science. Twenty-three volumes published between 1923-1956. New York: Columbia University Press.

Thurman, Howard. With Head and Heart. New York and London: Harcourt Brace Jovanovick, 1979.

Toulmin, Stephen. *From Form to Function: Philosophy and History of Science in the 1950s and Now.* Daedalus. Volume I, number 106, (Summer, 1977): 143-162.

Townsend, Harvey Gates. Philosophical Ideas in the United States. New York: American Book Company, 1934.

Turner, James. Without God, Without Creed. The Origins of Unbelief In America. Baltimore and London: Johns Hopkins Press, 1985.

Villemaire, Diane. *Metaphysics In the Twentieth Century: The Philosophies of History of R. G. Collingwood and E. A. Burtt.* Master's Thesis in History at the University of Vermont, 1994.

Walsh, Paul. *Some Metaphysical Assumptions In Dewey's Philosophy.* The Journal of Philosophy 51 (December 25, 1954): 861-67.

Welchman, Jennifer. Dewey's Ethical Thought. Ithaca, New York and London: Cornell University Press, 1995.

Westfall, Richard S. Science and Religion in Seventeenth-Century England. New Haven: Yale Press, 1958.

____. The Construction of Modern Science. Cambridge: Cambridge University Press, 1977.

Westman, Robert S. *Two Cultures or One? A Second Look at Kuhn's The Copernican Revolution.* Isis. 85 (1994): 79-115.

White, Morton. Pragmatism and the American Mind, Essays and Reviews in Philosophy and Intellectual History. New York: Oxford University Press, 1973.

____. Social Thought in America, The Revolt Against Formalism. Boston: Beacon Press, 1964.

____. The Origin of Dewey's Instrumentalism. New York: Columbia U. Press, 1943.

____. *The Revolt Against Formalism in American Social Thought of the Twentieth Century.* Journal of the History of Ideas. Volume VIII, number 2 (April, 1947): 131-152.

Wilson, Daniel J. Science, Community and the Transformation of American Philosophy, 1860-1930. Chicago and London: University of Chicago Press, 1990.

Wilson, Edwin H. The Genesis of a Humanist Manifesto. Amherst, NY.: Humanist Press, 1955. http://www.infidels.org/library/modern/edwin_wilson/manifesto/

Wilson, Margaret. Descartes. London: Routledge and Kegan Paul, 1987.

Winn, Ralph B., editor. American Philosophy. New York: Philosophical Library, Inc., 1955.

Wish, Harvey. *Introduction* to J. H. Robinson, The New History. First edition, 1912. New York: Macmillan Co., 1965.

Wood, John K. The Nature of Conflicts Between Science and Religion. Logan, Utah, Utah State University Press, Monograph Series, Volume IX, Number 2, (March, 1962).

Woodbridge, Frederick J. E. The Purpose of History. New York: Columbia University Press, 1916.

——' An Essay on Nature,. New York: Columbia University Press, 1940.

____. Nature and Mind: Selected Essays. New York: Russell and Russell, 1965.

INDEX

M

N

O

P

Q

R

S

Boston Studies in the Philosophy of Science

Editor: Robert S. Cohen, *Boston University*

1. M.W. Wartofsky (ed.): *Proceedings of the Boston Colloquium for the Philosophy of Science, 1961/1962.* [Synthese Library 6] 1963 ISBN 90-277-0021-4
2. R.S. Cohen and M.W. Wartofsky (eds.): *Proceedings of the Boston Colloquium for the Philosophy of Science, 1962/1964.* In Honor of P. Frank. [Synthese Library 10] 1965 ISBN 90-277-9004-0
3. R.S. Cohen and M.W. Wartofsky (eds.): *Proceedings of the Boston Colloquium for the Philosophy of Science, 1964/1966.* In Memory of Norwood Russell Hanson. [Synthese Library 14] 1967 ISBN 90-277-0013-3
4. R.S. Cohen and M.W. Wartofsky (eds.): *Proceedings of the Boston Colloquium for the Philosophy of Science, 1966/1968.* [Synthese Library 18] 1969 ISBN 90-277-0014-1
5. R.S. Cohen and M.W. Wartofsky (eds.): *Proceedings of the Boston Colloquium for the Philosophy of Science, 1966/1968.* [Synthese Library 19] 1969 ISBN 90-277-0015-X
6. R.S. Cohen and R.J. Seeger (eds.): *Ernst Mach, Physicist and Philosopher.* [Synthese Library 27] 1970 ISBN 90-277-0016-8
7. M. Čapek: *Bergson and Modern Physics.* A Reinterpretation and Re-evaluation. [Synthese Library 37] 1971 ISBN 90-277-0186-5
8. R.C. Buck and R.S. Cohen (eds.): *PSA 1970.* Proceedings of the 2nd Biennial Meeting of the Philosophy and Science Association (Boston, Fall 1970). In Memory of Rudolf Carnap. [Synthese Library 39] 1971 ISBN 90-277-0187-3; Pb 90-277-0309-4
9. A.A. Zinov'ev: *Foundations of the Logical Theory of Scientific Knowledge (Complex Logic).* Translated from Russian. Revised and enlarged English Edition, with an Appendix by G.A. Smirnov, E.A. Sidorenko, A.M. Fedina and L.A. Bobrova. [Synthese Library 46] 1973 ISBN 90-277-0193-8; Pb 90-277-0324-8
10. L. Tondl: *Scientific Procedures.* A Contribution Concerning the Methodological Problems of Scientific Concepts and Scientific Explanation.Translated from Czech. [Synthese Library 47] 1973 ISBN 90-277-0147-4; Pb 90-277-0323-X
11. R.J. Seeger and R.S. Cohen (eds.): *Philosophical Foundations of Science.* Proceedings of Section L, 1969, American Association for the Advancement of Science. [Synthese Library 58] 1974 ISBN 90-277-0390-6; Pb 90-277-0376-0
12. A. Grünbaum: *Philosophical Problems of Space and Times.* 2nd enlarged ed. [Synthese Library 55] 1973 ISBN 90-277-0357-4; Pb 90-277-0358-2
13. R.S. Cohen and M.W. Wartofsky (eds.): *Logical and Epistemological Studies in Contemporary Physics.* Proceedings of the Boston Colloquium for the Philosophy of Science, 1969/72, Part I. [Synthese Library 59] 1974 ISBN 90-277-0391-4; Pb 90-277-0377-9
14. R.S. Cohen and M.W. Wartofsky (eds.): *Methodological and Historical Essays in the Natural and Social Sciences.* Proceedings of the Boston Colloquium for the Philosophy of Science, 1969/72, Part II. [Synthese Library 60] 1974 ISBN 90-277-0392-2; Pb 90-277-0378-7
15. R.S. Cohen, J.J. Stachel and M.W. Wartofsky (eds.): *For Dirk Struik.* Scientific, Historical and Political Essays in Honor of Dirk J. Struik. [Synthese Library 61] 1974 ISBN 90-277-0393-0; Pb 90-277-0379-5
16. N. Geschwind: *Selected Papers on Language and the Brains.* [Synthese Library 68] 1974 ISBN 90-277-0262-4; Pb 90-277-0263-2
17. B.G. Kuznetsov: *Reason and Being.* Translated from Russian. Edited by C.R. Fawcett and R.S. Cohen. 1987 ISBN 90-277-2181-5

18. P. Mittelstaedt: *Philosophical Problems of Modern Physics.* Translated from the revised 4th German edition by W. Riemer and edited by R.S. Cohen. [Synthese Library 95] 1976
ISBN 90-277-0285-3; Pb 90-277-0506-2
19. H. Mehlberg: *Time, Causality, and the Quantum Theory.* Studies in the Philosophy of Science. Vol. I: *Essay on the Causal Theory of Time.* Vol. II: *Time in a Quantized Universe.* Translated from French. Edited by R.S. Cohen. 1980 Vol. I: ISBN 90-277-0721-9; Pb 90-277-1074-0
Vol. II: ISBN 90-277-1075-9; Pb 90-277-1076-7
20. K.F. Schaffner and R.S. Cohen (eds.): *PSA 1972.* Proceedings of the 3rd Biennial Meeting of the Philosophy of Science Association (Lansing, Michigan, Fall 1972). [Synthese Library 64] 1974 ISBN 90-277-0408-2; Pb 90-277-0409-0
21. R.S. Cohen and J.J. Stachel (eds.): *Selected Papers of Léon Rosenfeld.* [Synthese Library 100] 1979 ISBN 90-277-0651-4; Pb 90-277-0652-2
22. M. Čapek (ed.): *The Concepts of Space and Time.* Their Structure and Their Development. [Synthese Library 74] 1976 ISBN 90-277-0355-8; Pb 90-277-0375-2
23. M. Grene: *The Understanding of Nature.* Essays in the Philosophy of Biology. [Synthese Library 66] 1974 ISBN 90-277-0462-7; Pb 90-277-0463-5
24. D. Ihde: *Technics and Praxis.* A Philosophy of Technology. [Synthese Library 130] 1979
ISBN 90-277-0953-X; Pb 90-277-0954-8
25. J. Hintikka and U. Remes: *The Method of Analysis.* Its Geometrical Origin and Its General Significance. [Synthese Library 75] 1974 ISBN 90-277-0532-1; Pb 90-277-0543-7
26. J.E. Murdoch and E.D. Sylla (eds.): *The Cultural Context of Medieval Learning.* Proceedings of the First International Colloquium on Philosophy, Science, and Theology in the Middle Ages, 1973. [Synthese Library 76] 1975 ISBN 90-277-0560-7; Pb 90-277-0587-9
27. M. Grene and E. Mendelsohn (eds.): *Topics in the Philosophy of Biology.* [Synthese Library 84] 1976 ISBN 90-277-0595-X; Pb 90-277-0596-8
28. J. Agassi: *Science in Flux.* [Synthese Library 80] 1975
ISBN 90-277-0584-4; Pb 90-277-0612-3
29. J.J. Wiatr (ed.): *Polish Essays in the Methodology of the Social Sciences.* [Synthese Library 131] 1979 ISBN 90-277-0723-5; Pb 90-277-0956-4
30. P. Janich: *Protophysics of Time.* Constructive Foundation and History of Time Measurement. Translated from German. 1985 ISBN 90-277-0724-3
31. R.S. Cohen and M.W. Wartofsky (eds.): *Language, Logic, and Method.* 1983
ISBN 90-277-0725-1
32. R.S. Cohen, C.A. Hooker, A.C. Michalos and J.W. van Evra (eds.): *PSA 1974.* Proceedings of the 4th Biennial Meeting of the Philosophy of Science Association. [Synthese Library 101] 1976 ISBN 90-277-0647-6; Pb 90-277-0648-4
33. G. Holton and W.A. Blanpied (eds.): *Science and Its Public.* The Changing Relationship. [Synthese Library 96] 1976 ISBN 90-277-0657-3; Pb 90-277-0658-1
34. M.D. Grmek, R.S. Cohen and G. Cimino (eds.): *On Scientific Discovery.* The 1977 Erice Lectures. 1981 ISBN 90-277-1122-4; Pb 90-277-1123-2
35. S. Amsterdamski: *Between Experience and Metaphysics.* Philosophical Problems of the Evolution of Science. Translated from Polish. [Synthese Library 77] 1975
ISBN 90-277-0568-2; Pb 90-277-0580-1
36. M. Marković and G. Petrović (eds.): *Praxis.* Yugoslav Essays in the Philosophy and Methodology of the Social Sciences. [Synthese Library 134] 1979
ISBN 90-277-0727-8; Pb 90-277-0968-8

Boston Studies in the Philosophy of Science

37. H. von Helmholtz: *Epistemological Writings.* The Paul Hertz / Moritz Schlick Centenary Edition of 1921. Translated from German by M.F. Lowe. Edited with an Introduction and Bibliography by R.S. Cohen and Y. Elkana. [Synthese Library 79] 1977 ISBN 90-277-0290-X; Pb 90-277-0582-8
38. R.M. Martin: *Pragmatics, Truth and Language.* 1979 ISBN 90-277-0992-0; Pb 90-277-0993-9
39. R.S. Cohen, P.K. Feyerabend and M.W. Wartofsky (eds.): *Essays in Memory of Imre Lakatos.* [Synthese Library 99] 1976 ISBN 90-277-0654-9; Pb 90-277-0655-7
40. Not published.
41. Not published.
42. H.R. Maturana and F.J. Varela: *Autopoiesis and Cognition.* The Realization of the Living. With a Preface to "Autopoiesis' by S. Beer. 1980 ISBN 90-277-1015-5; Pb 90-277-1016-3
43. A. Kasher (ed.): *Language in Focus: Foundations, Methods and Systems.* Essays in Memory of Yehoshua Bar-Hillel. [Synthese Library 89] 1976 ISBN 90-277-0644-1; Pb 90-277-0645-X
44. T.D. Thao: *Investigations into the Origin of Language and Consciousness.* 1984 ISBN 90-277-0827-4
45. F.G.-I. Nagasaka (ed.): *Japanese Studies in the Philosophy of Science.* 1997 ISBN 0-7923-4781-1
46. P.L. Kapitza: *Experiment, Theory, Practice.* Articles and Addresses. Edited by R.S. Cohen. 1980 ISBN 90-277-1061-9; Pb 90-277-1062-7
47. M.L. Dalla Chiara (ed.): *Italian Studies in the Philosophy of Science.* 1981 ISBN 90-277-0735-9; Pb 90-277-1073-2
48. M.W. Wartofsky: *Models.* Representation and the Scientific Understanding. [Synthese Library 129] 1979 ISBN 90-277-0736-7; Pb 90-277-0947-5
49. T.D. Thao: *Phenomenology and Dialectical Materialism.* Edited by R.S. Cohen. 1986 ISBN 90-277-0737-5
50. Y. Fried and J. Agassi: *Paranoia.* A Study in Diagnosis. [Synthese Library 102] 1976 ISBN 90-277-0704-9; Pb 90-277-0705-7
51. K.H. Wolff: *Surrender and Cath.* Experience and Inquiry Today. [Synthese Library 105] 1976 ISBN 90-277-0758-8; Pb 90-277-0765-0
52. K. Kosík: *Dialectics of the Concrete.* A Study on Problems of Man and World. 1976 ISBN 90-277-0761-8; Pb 90-277-0764-2
53. N. Goodman: *The Structure of Appearance.* [Synthese Library 107] 1977 ISBN 90-277-0773-1; Pb 90-277-0774-X
54. H.A. Simon: *Models of Discovery* and Other Topics in the Methods of Science. [Synthese Library 114] 1977 ISBN 90-277-0812-6; Pb 90-277-0858-4
55. M. Lazerowitz: *The Language of Philosophy.* Freud and Wittgenstein. [Synthese Library 117] 1977 ISBN 90-277-0826-6; Pb 90-277-0862-2
56. T. Nickles (ed.): *Scientific Discovery, Logic, and Rationality.* 1980 ISBN 90-277-1069-4; Pb 90-277-1070-8
57. J. Margolis: *Persons and Mind.* The Prospects of Nonreductive Materialism. [Synthese Library 121] 1978 ISBN 90-277-0854-1; Pb 90-277-0863-0
58. G. Radnitzky and G. Andersson (eds.): *Progress and Rationality in Science.* [Synthese Library 125] 1978 ISBN 90-277-0921-1; Pb 90-277-0922-X
59. G. Radnitzky and G. Andersson (eds.): *The Structure and Development of Science.* [Synthese Library 136] 1979 ISBN 90-277-0994-7; Pb 90-277-0995-5

Boston Studies in the Philosophy of Science

60. T. Nickles (ed.): *Scientific Discovery.* Case Studies. 1980 ISBN 90-277-1092-9; Pb 90-277-1093-7
61. M.A. Finocchiaro: *Galileo and the Art of Reasoning.* Rhetorical Foundation of Logic and Scientific Method. 1980 ISBN 90-277-1094-5; Pb 90-277-1095-3
62. W.A. Wallace: *Prelude to Galileo.* Essays on Medieval and 16th-Century Sources of Galileo's Thought. 1981 ISBN 90-277-1215-8; Pb 90-277-1216-6
63. F. Rapp: *Analytical Philosophy of Technology.* Translated from German. 1981 ISBN 90-277-1221-2; Pb 90-277-1222-0
64. R.S. Cohen and M.W. Wartofsky (eds.): *Hegel and the Sciences.* 1984 ISBN 90-277-0726-X
65. J. Agassi: *Science and Society.* Studies in the Sociology of Science. 1981 ISBN 90-277-1244-1; Pb 90-277-1245-X
66. L. Tondl: *Problems of Semantics.* A Contribution to the Analysis of the Language of Science. Translated from Czech. 1981 ISBN 90-277-0148-2; Pb 90-277-0316-7
67. J. Agassi and R.S. Cohen (eds.): *Scientific Philosophy Today.* Essays in Honor of Mario Bunge. 1982 ISBN 90-277-1262-X; Pb 90-277-1263-8
68. W. Krajewski (ed.): *Polish Essays in the Philosophy of the Natural Sciences.* Translated from Polish and edited by R.S. Cohen and C.R. Fawcett. 1982 ISBN 90-277-1286-7; Pb 90-277-1287-5
69. J.H. Fetzer: *Scientific Knowledge.* Causation, Explanation and Corroboration. 1981 ISBN 90-277-1335-9; Pb 90-277-1336-7
70. S. Grossberg: *Studies of Mind and Brain.* Neural Principles of Learning, Perception, Development, Cognition, and Motor Control. 1982 ISBN 90-277-1359-6; Pb 90-277-1360-X
71. R.S. Cohen and M.W. Wartofsky (eds.): *Epistemology, Methodology, and the Social Sciences.* 1983. ISBN 90-277-1454-1
72. K. Berka: *Measurement.* Its Concepts, Theories and Problems. Translated from Czech. 1983 ISBN 90-277-1416-9
73. G.L. Pandit: *The Structure and Growth of Scientific Knowledge.* A Study in the Methodology of Epistemic Appraisal. 1983 ISBN 90-277-1434-7
74. A.A. Zinov'ev: *Logical Physics.* Translated from Russian. Edited by R.S. Cohen. 1983 [*see also* Volume 9] ISBN 90-277-0734-0
75. G-G. Granger: *Formal Thought and the Sciences of Man.* Translated from French. With and Introduction by A. Rosenberg. 1983 ISBN 90-277-1524-6
76. R.S. Cohen and L. Laudan (eds.): *Physics, Philosophy and Psychoanalysis.* Essays in Honor of Adolf Grünbaum. 1983 ISBN 90-277-1533-5
77. G. Böhme, W. van den Daele, R. Hohlfeld, W. Krohn and W. Schäfer: *Finalization in Science.* The Social Orientation of Scientific Progress. Translated from German. Edited by W. Schäfer. 1983 ISBN 90-277-1549-1
78. D. Shapere: *Reason and the Search for Knowledge.* Investigations in the Philosophy of Science. 1984 ISBN 90-277-1551-3; Pb 90-277-1641-2
79. G. Andersson (ed.): *Rationality in Science and Politics.* Translated from German. 1984 ISBN 90-277-1575-0; Pb 90-277-1953-5
80. P.T. Durbin and F. Rapp (eds.): *Philosophy and Technology.* [*Also* Philosophy and Technology Series, Vol. 1] 1983 ISBN 90-277-1576-9
81. M. Marković: *Dialectical Theory of Meaning.* Translated from Serbo-Croat. 1984 ISBN 90-277-1596-3
82. R.S. Cohen and M.W. Wartofsky (eds.): *Physical Sciences and History of Physics.* 1984. ISBN 90-277-1615-3

Boston Studies in the Philosophy of Science

83. É. Meyerson: *The Relativistic Deduction.* Epistemological Implications of the Theory of Relativity. Translated from French. With a Review by Albert Einstein and an Introduction by Milić Čapek. 1985 ISBN 90-277-1699-4
84. R.S. Cohen and M.W. Wartofsky (eds.): *Methodology, Metaphysics and the History of Science.* In Memory of Benjamin Nelson. 1984 ISBN 90-277-1711-7
85. G. Tamás: *The Logic of Categories.* Translated from Hungarian. Edited by R.S. Cohen. 1986 ISBN 90-277-1742-7
86. S.L. de C. Fernandes: *Foundations of Objective Knowledge.* The Relations of Popper's Theory of Knowledge to That of Kant. 1985 ISBN 90-277-1809-1
87. R.S. Cohen and T. Schnelle (eds.): *Cognition and Fact.* Materials on Ludwik Fleck. 1986 ISBN 90-277-1902-0
88. G. Freudenthal: *Atom and Individual in the Age of Newton.* On the Genesis of the Mechanistic World View. Translated from German. 1986 ISBN 90-277-1905-5
89. A. Donagan, A.N. Perovich Jr and M.V. Wedin (eds.): *Human Nature and Natural Knowledge.* Essays presented to Marjorie Grene on the Occasion of Her 75th Birthday. 1986 ISBN 90-277-1974-8
90. C. Mitcham and A. Hunning (eds.): *Philosophy and Technology II.* Information Technology and Computers in Theory and Practice. [*Also* Philosophy and Technology Series, Vol. 2] 1986 ISBN 90-277-1975-6
91. M. Grene and D. Nails (eds.): *Spinoza and the Sciences.* 1986 ISBN 90-277-1976-4
92. S.P. Turner: *The Search for a Methodology of Social Science.* Durkheim, Weber, and the 19th-Century Problem of Cause, Probability, and Action. 1986. ISBN 90-277-2067-3
93. I.C. Jarvie: *Thinking about Society.* Theory and Practice. 1986 ISBN 90-277-2068-1
94. E. Ullmann-Margalit (ed.): *The Kaleidoscope of Science.* The Israel Colloquium: Studies in History, Philosophy, and Sociology of Science, Vol. 1. 1986 ISBN 90-277-2158-0; Pb 90-277-2159-9
95. E. Ullmann-Margalit (ed.): *The Prism of Science.* The Israel Colloquium: Studies in History, Philosophy, and Sociology of Science, Vol. 2. 1986 ISBN 90-277-2160-2; Pb 90-277-2161-0
96. G. Márkus: *Language and Production.* A Critique of the Paradigms. Translated from French. 1986 ISBN 90-277-2169-6
97. F. Amrine, F.J. Zucker and H. Wheeler (eds.): *Goethe and the Sciences: A Reappraisal.* 1987 ISBN 90-277-2265-X; Pb 90-277-2400-8
98. J.C. Pitt and M. Pera (eds.): *Rational Changes in Science.* Essays on Scientific Reasoning. Translated from Italian. 1987 ISBN 90-277-2417-2
99. O. Costa de Beauregard: *Time, the Physical Magnitude.* 1987 ISBN 90-277-2444-X
100. A. Shimony and D. Nails (eds.): *Naturalistic Epistemology.* A Symposium of Two Decades. 1987 ISBN 90-277-2337-0
101. N. Rotenstreich: *Time and Meaning in History.* 1987 ISBN 90-277-2467-9
102. D.B. Zilberman: *The Birth of Meaning in Hindu Thought.* Edited by R.S. Cohen. 1988 ISBN 90-277-2497-0
103. T.F. Glick (ed.): *The Comparative Reception of Relativity.* 1987 ISBN 90-277-2498-9
104. Z. Harris, M. Gottfried, T. Ryckman, P. Mattick Jr, A. Daladier, T.N. Harris and S. Harris: *The Form of Information in Science.* Analysis of an Immunology Sublanguage. With a Preface by Hilary Putnam. 1989 ISBN 90-277-2516-0
105. F. Burwick (ed.): *Approaches to Organic Form.* Permutations in Science and Culture. 1987 ISBN 90-277-2541-1

Boston Studies in the Philosophy of Science

106. M. Almási: *The Philosophy of Appearances.* Translated from Hungarian. 1989 ISBN 90-277-2150-5
107. S. Hook, W.L. O'Neill and R. O'Toole (eds.): *Philosophy, History and Social Action.* Essays in Honor of Lewis Feuer. With an Autobiographical Essay by L. Feuer. 1988 ISBN 90-277-2644-2
108. I. Hronszky, M. Fehér and B. Dajka: *Scientific Knowledge Socialized.* Selected Proceedings of the 5th Joint International Conference on the History and Philosophy of Science organized by the IUHPS (Veszprém, Hungary, 1984). 1988 ISBN 90-277-2284-6
109. P. Tillers and E.D. Green (eds.): *Probability and Inference in the Law of Evidence.* The Uses and Limits of Bayesianism. 1988 ISBN 90-277-2689-2
110. E. Ullmann-Margalit (ed.): *Science in Reflection.* The Israel Colloquium: Studies in History, Philosophy, and Sociology of Science, Vol. 3. 1988 ISBN 90-277-2712-0; Pb 90-277-2713-9
111. K. Gavroglu, Y. Goudaroulis and P. Nicolacopoulos (eds.): *Imre Lakatos and Theories of Scientific Change.* 1989 ISBN 90-277-2766-X
112. B. Glassner and J.D. Moreno (eds.): *The Qualitative-Quantitative Distinction in the Social Sciences.* 1989 ISBN 90-277-2829-1
113. K. Arens: *Structures of Knowing.* Psychologies of the 19th Century. 1989 ISBN 0-7923-0009-2
114. A. Janik: *Style, Politics and the Future of Philosophy.* 1989 ISBN 0-7923-0056-4
115. F. Amrine (ed.): *Literature and Science as Modes of Expression.* With an Introduction by S. Weininger. 1989 ISBN 0-7923-0133-1
116. J.R. Brown and J. Mittelstrass (eds.): *An Intimate Relation.* Studies in the History and Philosophy of Science. Presented to Robert E. Butts on His 60th Birthday. 1989 ISBN 0-7923-0169-2
117. F. D'Agostino and I.C. Jarvie (eds.): *Freedom and Rationality.* Essays in Honor of John Watkins. 1989 ISBN 0-7923-0264-8
118. D. Zolo: *Reflexive Epistemology.* The Philosophical Legacy of Otto Neurath. 1989 ISBN 0-7923-0320-2
119. M. Kearn, B.S. Philips and R.S. Cohen (eds.): *Georg Simmel and Contemporary Sociology.* 1989 ISBN 0-7923-0407-1
120. T.H. Levere and W.R. Shea (eds.): *Nature, Experiment and the Science.* Essays on Galileo and the Nature of Science. In Honour of Stillman Drake. 1989 ISBN 0-7923-0420-9
121. P. Nicolacopoulos (ed.): *Greek Studies in the Philosophy and History of Science.* 1990 ISBN 0-7923-0717-8
122. R. Cooke and D. Costantini (eds.): *Statistics in Science.* The Foundations of Statistical Methods in Biology, Physics and Economics. 1990 ISBN 0-7923-0797-6
123. P. Duhem: *The Origins of Statics.* Translated from French by G.F. Leneaux, V.N. Vagliente and G.H. Wagner. With an Introduction by S.L. Jaki. 1991 ISBN 0-7923-0898-0
124. H. Kamerlingh Onnes: *Through Measurement to Knowledge.* The Selected Papers, 1853-1926. Edited and with an Introduction by K. Gavroglu and Y. Goudaroulis. 1991 ISBN 0-7923-0825-5
125. M. Čapek: *The New Aspects of Time: Its Continuity and Novelties.* Selected Papers in the Philosophy of Science. 1991 ISBN 0-7923-0911-1
126. S. Unguru (ed.): *Physics, Cosmology and Astronomy, 1300–1700.* Tension and Accommodation. 1991 ISBN 0-7923-1022-5

Boston Studies in the Philosophy of Science

127. Z. Bechler: *Newton's Physics on the Conceptual Structure of the Scientific Revolution.* 1991
ISBN 0-7923-1054-3
128. É. Meyerson: *Explanation in the Sciences.* Translated from French by M-A. Siple and D.A. Siple. 1991 ISBN 0-7923-1129-9
129. A.I. Tauber (ed.): *Organism and the Origins of Self.* 1991 ISBN 0-7923-1185-X
130. F.J. Varela and J-P. Dupuy (eds.): *Understanding Origins.* Contemporary Views on the Origin of Life, Mind and Society. 1992 ISBN 0-7923-1251-1
131. G.L. Pandit: *Methodological Variance.* Essays in Epistemological Ontology and the Methodology of Science. 1991 ISBN 0-7923-1263-5
132. G. Munévar (ed.): *Beyond Reason.* Essays on the Philosophy of Paul Feyerabend. 1991
ISBN 0-7923-1272-4
133. T.E. Uebel (ed.): *Rediscovering the Forgotten Vienna Circle.* Austrian Studies on Otto Neurath and the Vienna Circle. Partly translated from German. 1991 ISBN 0-7923-1276-7
134. W.R. Woodward and R.S. Cohen (eds.): *World Views and Scientific Discipline Formation.* Science Studies in the [former] German Democratic Republic. Partly translated from German by W.R. Woodward. 1991 ISBN 0-7923-1286-4
135. P. Zambelli: *The Speculum Astronomiae and Its Enigma.* Astrology, Theology and Science in Albertus Magnus and His Contemporaries. 1992 ISBN 0-7923-1380-1
136. P. Petitjean, C. Jami and A.M. Moulin (eds.): *Science and Empires.* Historical Studies about Scientific Development and European Expansion. ISBN 0-7923-1518-9
137. W.A. Wallace: *Galileo's Logic of Discovery and Proof.* The Background, Content, and Use of His Appropriated Treatises on Aristotle's *Posterior Analytics.* 1992 ISBN 0-7923-1577-4
138. W.A. Wallace: *Galileo's Logical Treatises.* A Translation, with Notes and Commentary, of His Appropriated Latin Questions on Aristotle's *Posterior Analytics.* 1992 ISBN 0-7923-1578-2
Set (137 + 138) ISBN 0-7923-1579-0
139. M.J. Nye, J.L. Richards and R.H. Stuewer (eds.): *The Invention of Physical Science.* Intersections of Mathematics, Theology and Natural Philosophy since the Seventeenth Century. Essays in Honor of Erwin N. Hiebert. 1992 ISBN 0-7923-1753-X
140. G. Corsi, M.L. dalla Chiara and G.C. Ghirardi (eds.): *Bridging the Gap: Philosophy, Mathematics and Physics.* Lectures on the Foundations of Science. 1992 ISBN 0-7923-1761-0
141. C.-H. Lin and D. Fu (eds.): *Philosophy and Conceptual History of Science in Taiwan.* 1992
ISBN 0-7923-1766-1
142. S. Sarkar (ed.): *The Founders of Evolutionary Genetics.* A Centenary Reappraisal. 1992
ISBN 0-7923-1777-7
143. J. Blackmore (ed.): *Ernst Mach – A Deeper Look.* Documents and New Perspectives. 1992
ISBN 0-7923-1853-6
144. P. Kroes and M. Bakker (eds.): *Technological Development and Science in the Industrial Age.* New Perspectives on the Science–Technology Relationship. 1992 ISBN 0-7923-1898-6
145. S. Amsterdamski: *Between History and Method.* Disputes about the Rationality of Science. 1992 ISBN 0-7923-1941-9
146. E. Ullmann-Margalit (ed.): *The Scientific Enterprise.* The Bar-Hillel Colloquium: Studies in History, Philosophy, and Sociology of Science, Volume 4. 1992 ISBN 0-7923-1992-3
147. L. Embree (ed.): *Metaarchaeology.* Reflections by Archaeologists and Philosophers. 1992
ISBN 0-7923-2023-9
148. S. French and H. Kamminga (eds.): *Correspondence, Invariance and Heuristics.* Essays in Honour of Heinz Post. 1993 ISBN 0-7923-2085-9
149. M. Bunzl: *The Context of Explanation.* 1993 ISBN 0-7923-2153-7

Boston Studies in the Philosophy of Science

150. I.B. Cohen (ed.): *The Natural Sciences and the Social Sciences.* Some Critical and Historical Perspectives. 1994 ISBN 0-7923-2223-1
151. K. Gavroglu, Y. Christianidis and E. Nicolaidis (eds.): *Trends in the Historiography of Science.* 1994 ISBN 0-7923-2255-X
152. S. Poggi and M. Bossi (eds.): *Romanticism in Science.* Science in Europe, 1790–1840. 1994 ISBN 0-7923-2336-X
153. J. Faye and H.J. Folse (eds.): *Niels Bohr and Contemporary Philosophy.* 1994 ISBN 0-7923-2378-5
154. C.C. Gould and R.S. Cohen (eds.): *Artifacts, Representations, and Social Practice.* Essays for Marx W. Wartofsky. 1994 ISBN 0-7923-2481-1
155. R.E. Butts: *Historical Pragmatics.* Philosophical Essays. 1993 ISBN 0-7923-2498-6
156. R. Rashed: *The Development of Arabic Mathematics: Between Arithmetic and Algebra.* Translated from French by A.F.W. Armstrong. 1994 ISBN 0-7923-2565-6
157. I. Szumilewicz-Lachman (ed.): *Zygmunt Zawirski: His Life and Work.* With Selected Writings on Time, Logic and the Methodology of Science. Translations by Feliks Lachman. Ed. by R.S. Cohen, with the assistance of B. Bergo. 1994 ISBN 0-7923-2566-4
158. S.N. Haq: *Names, Natures and Things.* The Alchemist Jābir ibn Ḥayyān and His *Kitāb al-Aḥjār* (Book of Stones). 1994 ISBN 0-7923-2587-7
159. P. Plaass: *Kant's Theory of Natural Science.* Translation, Analytic Introduction and Commentary by Alfred E. and Maria G. Miller. 1994 ISBN 0-7923-2750-0
160. J. Misiek (ed.): *The Problem of Rationality in Science and its Philosophy.* On Popper vs. Polanyi. The Polish Conferences 1988–89. 1995 ISBN 0-7923-2925-2
161. I.C. Jarvie and N. Laor (eds.): *Critical Rationalism, Metaphysics and Science.* Essays for Joseph Agassi, Volume I. 1995 ISBN 0-7923-2960-0
162. I.C. Jarvie and N. Laor (eds.): *Critical Rationalism, the Social Sciences and the Humanities.* Essays for Joseph Agassi, Volume II. 1995 ISBN 0-7923-2961-9
Set (161–162) ISBN 0-7923-2962-7
163. K. Gavroglu, J. Stachel and M.W. Wartofsky (eds.): *Physics, Philosophy, and the Scientific Community.* Essays in the Philosophy and History of the Natural Sciences and Mathematics. In Honor of Robert S. Cohen. 1995 ISBN 0-7923-2988-0
164. K. Gavroglu, J. Stachel and M.W. Wartofsky (eds.): *Science, Politics and Social Practice.* Essays on Marxism and Science, Philosophy of Culture and the Social Sciences. In Honor of Robert S. Cohen. 1995 ISBN 0-7923-2989-9
165. K. Gavroglu, J. Stachel and M.W. Wartofsky (eds.): *Science, Mind and Art.* Essays on Science and the Humanistic Understanding in Art, Epistemology, Religion and Ethics. Essays in Honor of Robert S. Cohen. 1995 ISBN 0-7923-2990-2
Set (163–165) ISBN 0-7923-2991-0
166. K.H. Wolff: *Transformation in the Writing.* A Case of Surrender-and-Catch. 1995 ISBN 0-7923-3178-8
167. A.J. Kox and D.M. Siegel (eds.): *No Truth Except in the Details.* Essays in Honor of Martin J. Klein. 1995 ISBN 0-7923-3195-8
168. J. Blackmore: *Ludwig Boltzmann, His Later Life and Philosophy, 1900–1906.* Book One: A Documentary History. 1995 ISBN 0-7923-3231-8
169. R.S. Cohen, R. Hilpinen and R. Qiu (eds.): *Realism and Anti-Realism in the Philosophy of Science.* Beijing International Conference, 1992. 1996 ISBN 0-7923-3233-4
170. I. Kuçuradi and R.S. Cohen (eds.): *The Concept of Knowledge.* The Ankara Seminar. 1995 ISBN 0-7923-3241-5

Boston Studies in the Philosophy of Science

171. M.A. Grodin (ed.): *Meta Medical Ethics*: The Philosophical Foundations of Bioethics. 1995 ISBN 0-7923-3344-6
172. S. Ramirez and R.S. Cohen (eds.): *Mexican Studies in the History and Philosophy of Science.* 1995 ISBN 0-7923-3462-0
173. C. Dilworth: *The Metaphysics of Science.* An Account of Modern Science in Terms of Principles, Laws and Theories. 1995 ISBN 0-7923-3693-3
174. J. Blackmore: *Ludwig Boltzmann, His Later Life and Philosophy, 1900–1906* Book Two: The Philosopher. 1995 ISBN 0-7923-3464-7
175. P. Damerow: *Abstraction and Representation.* Essays on the Cultural Evolution of Thinking. 1996 ISBN 0-7923-3816-2
176. M.S. Macrakis: *Scarcity's Ways: The Origins of Capital.* A Critical Essay on Thermodynamics, Statistical Mechanics and Economics. 1997 ISBN 0-7923-4760-9
177. M. Marion and R.S. Cohen (eds.): *Québec Studies in the Philosophy of Science.* Part I: Logic, Mathematics, Physics and History of Science. Essays in Honor of Hugues Leblanc. 1995 ISBN 0-7923-3559-7
178. M. Marion and R.S. Cohen (eds.): *Québec Studies in the Philosophy of Science.* Part II: Biology, Psychology, Cognitive Science and Economics. Essays in Honor of Hugues Leblanc. 1996 ISBN 0-7923-3560-0
     Set (177–178) ISBN 0-7923-3561-9
179. Fan Dainian and R.S. Cohen (eds.): *Chinese Studies in the History and Philosophy of Science and Technology.* 1996 ISBN 0-7923-3463-9
180. P. Forman and J.M. Sánchez-Ron (eds.): *National Military Establishments and the Advancement of Science and Technology.* Studies in 20th Century History. 1996 ISBN 0-7923-3541-4
181. E.J. Post: *Quantum Reprogramming.* Ensembles and Single Systems: A Two-Tier Approach to Quantum Mechanics. 1995 ISBN 0-7923-3565-1
182. A.I. Tauber (ed.): *The Elusive Synthesis: Aesthetics and Science.* 1996 ISBN 0-7923-3904-5
183. S. Sarkar (ed.): *The Philosophy and History of Molecular Biology: New Perspectives.* 1996 ISBN 0-7923-3947-9
184. J.T. Cushing, A. Fine and S. Goldstein (eds.): *Bohmian Mechanics and Quantum Theory: An Appraisal.* 1996 ISBN 0-7923-4028-0
185. K. Michalski: *Logic and Time.* An Essay on Husserl's Theory of Meaning. 1996 ISBN 0-7923-4082-5
186. G. Munévar (ed.): *Spanish Studies in the Philosophy of Science.* 1996 ISBN 0-7923-4147-3
187. G. Schubring (ed.): *Hermann Günther Graßmann (1809–1877): Visionary Mathematician, Scientist and Neohumanist Scholar.* Papers from a Sesquicentennial Conference. 1996 ISBN 0-7923-4261-5
188. M. Bitbol: *Schrödinger's Philosophy of Quantum Mechanics.* 1996 ISBN 0-7923-4266-6
189. J. Faye, U. Scheffler and M. Urchs (eds.): *Perspectives on Time.* 1997 ISBN 0-7923-4330-1
190. K. Lehrer and J.C. Marek (eds.): *Austrian Philosophy Past and Present.* Essays in Honor of Rudolf Haller. 1996 ISBN 0-7923-4347-6
191. J.L. Lagrange: *Analytical Mechanics.* Translated and edited by Auguste Boissonade and Victor N. Vagliente. Translated from the *Mécanique Analytique, novelle édition* of 1811. 1997 ISBN 0-7923-4349-2
192. D. Ginev and R.S. Cohen (eds.): *Issues and Images in the Philosophy of Science.* Scientific and Philosophical Essays in Honour of Azarya Polikarov. 1997 ISBN 0-7923-4444-8

Boston Studies in the Philosophy of Science

193. R.S. Cohen, M. Horne and J. Stachel (eds.): *Experimental Metaphysics*. Quantum Mechanical Studies for Abner Shimony, Volume One. 1997 ISBN 0-7923-4452-9
194. R.S. Cohen, M. Horne and J. Stachel (eds.): *Potentiality, Entanglement and Passion-at-a-Distance*. Quantum Mechanical Studies for Abner Shimony, Volume Two. 1997 ISBN 0-7923-4453-7; Set 0-7923-4454-5
195. R.S. Cohen and A.I. Tauber (eds.): *Philosophies of Nature: The Human Dimension*. 1997 ISBN 0-7923-4579-7
196. M. Otte and M. Panza (eds.): *Analysis and Synthesis in Mathematics*. History and Philosophy. 1997 ISBN 0-7923-4570-3
197. A. Denkel: *The Natural Background of Meaning*. 1999 ISBN 0-7923-5331-5
198. D. Baird, R.I.G. Hughes and A. Nordmann (eds.): *Heinrich Hertz: Classical Physicist, Modern Philosopher*. 1999 ISBN 0-7923-4653-X
199. A. Franklin: *Can That be Right?* Essays on Experiment, Evidence, and Science. 1999 ISBN 0-7923-5464-8
200. D. Raven, W. Krohn and R.S. Cohen (eds.): *The Social Origins of Modern Science*. 2000 ISBN 0-7923-6457-0
201. Reserved
202. Reserved
203. B. Babich and R.S. Cohen (eds.): *Nietzsche, Theories of Knowledge, and Critical Theory*. Nietzsche and the Sciences I. 1999 ISBN 0-7923-5742-6
204. B. Babich and R.S. Cohen (eds.): *Nietzsche, Epistemology, and Philosophy of Science*. Nietzsche and the Science II. 1999 ISBN 0-7923-5743-4
205. R. Hooykaas: *Fact, Faith and Fiction in the Development of Science*. The Gifford Lectures given in the University of St Andrews 1976. 1999 ISBN 0-7923-5774-4
206. M. Fehér, O. Kiss and L. Ropolyi (eds.): *Hermeneutics and Science*. 1999 ISBN 0-7923-5798-1
207. R.M. MacLeod (ed.): *Science and the Pacific War*. Science and Survival in the Pacific, 1939-1945. 1999 ISBN 0-7923-5851-1
208. I. Hanzel: *The Concept of Scientific Law in the Philosophy of Science and Epistemology*. A Study of Theoretical Reason. 1999 ISBN 0-7923-5852-X
209. G. Helm; R.J. Deltete (ed./transl.): *The Historical Development of Energetics*. 1999 ISBN 0-7923-5874-0
210. A. Orenstein and P. Kotatko (eds.): *Knowledge, Language and Logic*. Questions for Quine. 1999 ISBN 0-7923-5986-0
211. R.S. Cohen and H. Levine (eds.): *Maimonides and the Sciences*. 2000 ISBN 0-7923-6053-2
212. H. Gourko, D.I. Williamson and A.I. Tauber (eds.): *The Evolutionary Biology Papers of Elie Metchnikoff*. 2000 ISBN 0-7923-6067-2
213. S. D'Agostino: *A History of the Ideas of Theoretical Physics*. Essays on the Nineteenth and Twentieth Century Physics. 2000 ISBN 0-7923-6094-X
214. S. Lelas: *Science and Modernity*. Toward An Integral Theory of Science. 2000 ISBN 0-7923-6303-5
215. E. Agazzi and M. Pauri (eds.): *The Reality of the Unobservable*. Observability, Unobservability and Their Impact on the Issue of Scientific Realism. 2000 ISBN 0-7923-6311-6
216. P. Hoyningen-Huene and H. Sankey (eds.): *Incommensurability and Related Matters*. 2001 ISBN 0-7923-6989-0
217. A. Nieto-Galan: *Colouring Textiles*. A History of Natural Dyestuffs in Industrial Europe. 2001 ISBN 0-7923-7022-8

Boston Studies in the Philosophy of Science

218. J. Blackmore, R. Itagaki and S. Tanaka (eds.): *Ernst Mach's Vienna 1895–1930.* Or Phenomenalism as Philosophy of Science. 2001 ISBN 0-7923-7122-4
219. R. Vihalemm (ed.): *Estonian Studies in the History and Philosophy of Science.* 2001 ISBN 0-7923-7189-5
220. W. Lefèvre (ed.): *Between Leibniz, Newton, and Kant.* Philosophy and Science in the Eighteenth Century. 2001 ISBN 0-7923-7198-4
221. T.F. Glick, M.Á. Puig-Samper and R. Ruiz (eds.): *The Reception of Darwinism in the Iberian World.* Spain, Spanish America and Brazil. 2001 ISBN 1-4020-0082-0
222. U. Klein (ed.): *Tools and Modes of Representation in the Laboratory Sciences.* 2001 ISBN 1-4020-0100-2
223. P. Duhem: *Mixture and Chemical Combination.* And Related Essays. Edited and translated, with an introduction, by Paul Needham. 2002 ISBN 1-4020-0232-7
224. J.C. Boudri: *What was Mechanical about Mechanics.* The Concept of Force Betweem Metaphysics and Mechanics from Newton to Lagrange. 2002 ISBN 1-4020-0233-5
225. B.E. Babich (ed.): *Hermeneutic Philosophy of Science, Van Gogh's Eyes, and God.* Essays in Honor of Patrick A. Heelan, S.J. 2002 ISBN 1-4020-0234-3
226. D. Davies Villemaire: E.A. Burtt, Historian and Philosopher. A Study of the author of The Metaphysical Foundations of Modern Physical Science. 2002 ISBN 1-4020-0428-1
227. L.J. Cohen: *Knowledge and Language.* Selected Essays of L. Jonathan Cohen. Edited and with an introduction by James Logue. 2002 ISBN 1-4020-0474-5
228. G.E. Allen and R.M. MacLeod (eds.): *Science, History and Social Activism: A Tribute to Everett Mendelsohn.* 2002 ISBN 1-4020-0495-0

Also of interest:

R.S. Cohen and M.W. Wartofsky (eds.): *A Portrait of Twenty-Five Years Boston Colloquia for the Philosophy of Science, 1960-1985.* 1985 ISBN Pb 90-277-1971-3

Previous volumes are still available.

KLUWER ACADEMIC PUBLISHERS – DORDRECHT / BOSTON / LONDON

MIX
Papier aus verantwortungsvollen Quellen
Paper from responsible sources
FSC® C105338

If you have any concerns about our products,
you can contact us on
ProductSafety@springernature.com

In case Publisher is established outside the EU,
the EU authorized representative is:
Springer Nature Customer Service Center GmbH
Europaplatz 3, 69115 Heidelberg, Germany

Printed by Libri Plureos GmbH
in Hamburg, Germany